EDUCATION OF THE GIFTED IN EUROPE

EUROPEAN MEETINGS ON EDUCATIONAL RESEARCH

Part A:
Reports of Educational Research Symposia, Colloquies, and Workshops organised under the auspices of the Council of Europe

Part B:
European Conferences of Directors of Educational Research Institutions organised in cooperation with UNESCO, the UNESCO Institute for Education in Hamburg, and the Council of Europe

PART A: VOLUME 28

Part B Volume 23	The effectiveness of In-Service Education and Training of Teachers and School Leaders
Part A Volume 24	The Evaluation of Educational Programmes: Methods, Uses and Benefits
Part A Volume 25	Assessment of Pupil Achievement: Motivation and School Success
Part A Volume 26	Socialisation of Schoolchildren and their Education for Democratic Values and Human Rights
Part A Volume 27	History and Social Studies – Methodologies of Textbook Analysis

COUNCIL OF EUROPE – STRASBOURG
CENTER FOR THE STUDY OF GIFTEDNESS – NIJMEGEN

Education of the Gifted in Europe: Theoretical and Research Issues

REPORT OF THE EDUCATIONAL RESEARCH WORKSHOP
HELD IN NIJMEGEN (THE NETHERLANDS)
23-26 JULY 1991

Edited by
Franz J. Mönks
Center for the Study of Giftedness, The Netherlands
Michael W. Katzko
University of Nijmegen, The Netherlands
Herman W. van Boxtel
University of Utrecht, The Netherlands

SWETS & ZEITLINGER B.V. AMSTERDAM / LISSE

PUBLISHING SERVICE

Library of Congress Cataloging-in-Publication Data

Education of the gifted in Europe: theoretical and research issues : report of the educational research workshop held in Nijmegen (The Netherlands) 23-26 July 1991 / edited by Franz J. Mönks, Michael W. Katzko & Herman W. van Boxtel.
p. cm. -- (European meetings on educational research ; v. 28. Part A, Reports of educational research symposia, colloquies, and workshops organised under the auspices of the Council of Europe)
At head of title: Council of Europe--Strasbourg, Center for the Study of Giftedness--Nijmegen
Includes bibliographical references.
ISBN 90 265 1262 7
1. Gifted children--Education--Europe--Congresses. I. Mönks, Franz J. 1932- II. Katzko, Michael W. 1949- III. Boxtel, Herman W. van, 1951- . IV. Center for the Study of Giftedness (Nijmegen, The Netherlands) V Series: European meetings on educational research ; v. 28. VI. Series: European meetings on educational research ; Part A, Reports of educational research symposia, colloquies, and workshops organised under the auspices of the Council of Europe
LC3997. A1E38 1992
371.95'094--dc20 92-5639
CIP

CIP-gegevens Koninklijke Bibliotheek, Den Haag

Education

Education of the gifted in Europe: theoretical and research issues : report of the educational research workshop held in Nijmegen (The Netherlands) 23-26 July 1991 / edited by Franz J. Mönks, Michael W. Katzko & Herman W. van Boxtel. – Amsterdam [etc]: Swets & Zeitlinger. (European meetings on educational research ; ISSN 0924-0578) vol. 28. Pt. A)
– Met lit. opg.
ISBN 90-265-1262-7
NUGI 724
Trefw.: onderwijs ; begaafde kinderen ; Europa.

Printed in the Netherlands by Offsetdrukkerij Kanters B.V., Alblasserdam
Cover design: Rob Molthoff
Cover printed in the Netherlands by Casparie IJsselstein

ISBN 90 265 1262 7
NUGI 724

Table of Contents **page**

Foreword

Michael Vorbeck
Head of the Section for Educational Research and Documentation,
Council of Europe
Strasbourg, France

The Educational Research Workshop on gifted children and adolescents, which took place in Nijmegen, was jointly organised by the Centre for the Study of Giftedness at the University of Nijmegen, the Council for Cultural Co-operation of the Council of Europe, the Dutch and German Ministries of Education, as well as the Institute for Educational Research in the Netherlands (SVO).

These educational research meetings regularly bring together research workers and administrators from a selected number of the 30 countries taking part in the work of the Council for Cultural Co-operation. The purpose is to compare research findings and to encourage co-operation in a field of particular interest.

The aims of the Workshop on gifted children and adolescents were:

- To identify and discuss research on gifted children and adolescents and how best to promote their talents.
- To define neglected areas of research in this field.
- To draw conclusions from research so far available.
- To sound possibilities of co-operation among participating research workers and the institutes.

The Council of Europe decided to organise the Workshop in order to tackle a field which has for a long time been a neglected area of research. The large number of documents and the high participation proved that interest in the research on giftedness is enormous.

The following countries were represented: Austria, Azerbaijan, Czechoslovakia, Denmark, Estonia, Finland, France, Germany, Italy, Latvia, Netherlands, Poland, Portugal, Romania, Russia, Slovenia, Spain, Sweden, Switzerland, Turkey, United Kingdom, Yugoslavia.

(Excused: Belgium, Bulgaria, Hungary.)

There were also observers from the Parliamentary Assembly and UNESCO, Indonesia, Peru, as well as non-governmental organisations such as the World Council for Gifted and Talented Children (WCGT), European Council for High Ability (ECHA), EUROTALENT, MENSA, Association Nationale Pour les Enfants Intellectuellement Précoces (ANPEIP), National Association for Curriculum Enrichment & Extension (NACE), National Association for Gifted Children (NAGC), Asociación para el desarrollo de la Creatividad y el Talento (CREDEYTA) and PHAROS.

The list of participants is given at the end of this book.

Seven commissioned papers (covering Germany, the Netherlands, Poland, Spain, Sweden, Switzerland and the United Kingdom) were presented in plenary sessions and then discussed in three working groups. National and individual reports from various countries, as well as lists of research projects and bibliographies, were tabled as background material. On the final day, the Rapporteur General, Drs. Herman W. van Boxtel from the Netherlands, summed op the situation and the conclusions emerging from the Workshop.

The Council of Europe is particularly grateful to Professor Mönks for his excellent work in preparing and organising the Workshop. The Council of Europe would also like to express its thanks to the Rapporteur General, Drs. Herman W. van Boxtel, to the lecturers and to the group chairmen and rapporteurs. The English editing was done by Dr. Michael Katzko.

Editors' Preface

This book includes the main results of the Educational Research Workshop on gifted children and adolescents held in Nijmegen in 1991.

The book arranges the contributions to the Workshop into four Parts. Part I contains a General Introduction by the organizer and chairman of the Workshop, followed by the Final Report of the Workshop prepared by the Rapporteur General. Part II contains the papers commissioned for the Workshop, and are arranged alphabetically by author. Part III contains a selection of the papers which were tabled at the Workshop as background material. We applied two criteria in making our selection. First, we included papers which provided an overview of education and/or research on giftedness in the country of the author's origin. These papers are intended to cover the same topics addressed in the commissioned papers. Second, we also included some papers as examples of the kinds of research now being carried out in various European countries. All chapters in Part III are also arranged alphabetically by author. Part IV includes two special contributions. The paper by John Raven had been sent to the Workshop as background material. Because of its scope and content, we thought his contribution was of general value for the field of giftedness and served as a fitting closing overview of the topic. Finally, we asked Ernst Hany to examine the problem of conducting a literature search on the topic of giftedness. His contribution can facilitate the work which is essential to research but which is often very time consuming.

We would like to thank all the contributors for their prompt reactions to our requests, and for the confidence which they gave to us. In performing our task as editors, it was necessary to make some modifications to the original manuscripts, not only in "standardizing" the English of this truly international gathering, but also as a result of the need to shorten some of the manuscripts for publication in this book. We hope that, in both cases, we have remained faithful to what the authors had intended to communicate. We also did our best to assure that the reference lists were complete and accurate; for various reasons, it was not always possible to achieve this goal. We hope that, in the latter instances, the reader will not be greatly inconvenienced.

We would also like to thank Marcella Mönks for formatting the text. Finally, we wish to express our gratitude to the Dutch Ministry of Education and Sciences for their financial support of the preparation of this book.

Nijmegen, February 1992

F.J.M
M.W.K
H.W.v.B

PART 1

General Introduction

From conception to realization

Franz J. Mönks
The Netherlands

FIRST STEPS

The first conception of a heart pacemaker was formulated in the year 1928. Another 32 years had to pass before this conception was realized in a practical implementation. Similar time lapses between conception and realization can be found with other inventions, e.g. first conception of electrophotography in 1937 and first implementation in 1959 (22 years), and for oral contraceptives the time lapse is 9 years (resp. 1951-1960). However, between conception and implementaion of the video tape recorder there is a time lapse of only 6 years (1950-1956)! How long will it take before our "findings" in the field of giftedness will lead to a concrete solution with a substantial impact on the lives of gifted children?

Of course, when we are dealing with inventions there is usually a clearly circumscribed area of application. When we are dealing with educational or psychological matters such areas of application cannot be so easily defined. Yet, the gifted child as well as the learning disabled child are educational realities we have to deal with. "To deal with" means that we have to find the best possible educational as well as instructional methods to meet the special needs of these children. The approach that is very often taken in practice is to treat unequals with equal measures to make them more equal.

Gifted children and prodigies do not need more or a better treatment, but a different treatment. A differential curriculum, however, requires identification of intellectual as well as social aptitudes. What makes a person gifted and what treatment is required, once we have identified that person?

Academic interest in the gifted

a) Révész (1887-1955) - This Dutch (Hungarian by origin) Psychologist (residing in Amsterdam after 1921) had published on the topic of giftedness as early as 1921. His famous monograph entitled "Psychology of a musical prodigy" appeared in 1925. His scientific interest in the gifted child, especially giftedness in music and performing arts, was more than episodic. In 1946 he published "Creative giftedness" and in 1952 "Talent and genius: Characteristics of a psychology of giftedness".

b) Waterink (1890-1966) - Another Dutch author was Waterink. In 1938 he published an article entitled "Giftedness as a qualification of the personality". In this article he emphasized that the behavioral correlates of giftedness were originality and productivity, which becomes apparent in outstanding achievements. He also pointed out that the realization of giftedness is always dependent on the value which the society/culture gives to outstanding achievements. Waterink was convinced that societal appreciation of high achievement is important but not sufficient. Intrinsic motivation was the main source of gifted behavior. Thus, Waterink's views on giftedness anticipate recent definitions. The gifted person was, in his view, a person who possesses high intellectual capacities, shows productive and original behavior, has high intrinsic motivation and needs appreciation by the environment.

c) Duminy (1959) - Duminy, one of Waterink's students, defended a dissertation with the title "Giftedness - Some psychological and pedagogical aspects". The main results of this dissertation were later published in 1965. Duminy discussed and evaluated some instructional and educational steps to be taken for an appropriate training of the gifted child. His dissertation and article were written in Africaans and hinted of superciliousness in some statements such as his proposition IV: "for the reason that the primitive Bantu experiences the cosmic relations in a total different way, he is maladjusted in a Western society, and for that reason, *from a Western perspective*, he can hardly be regarded as gifted" (Duminy, 1959; attached proposition IV, p. 190). Based on the assumption that the identification of giftedness can only occur with tests derived from a Western culture, Duminy's proposition would be true. But this is not what he is concerned about. He stated that by their very nature such differences imply that gifted people do not exist in non-Western cultures.

d) De Groot (1956) - The impact of A. de Groot on the IQ debate in the Netherlands was very substantial. A more detailed account of De Groot's contribution to the field of giftedness can be found in Span's contribution to this volume.

These contributions (a - d) can be regarded as the forerunners of scientific interest in giftedness in the Netherlands.

EMERGING INTEREST

In 1963 the author of the present chapter wrote a review article entitled "Contributions to the study of giftedness in childhood and adolescence". In this article a survey was made of the current trends in the study of gifted children and adolescents and was published in a German and a Dutch psychological journal. In a thorough literature review six theoretical approaches were identified and can be summarized as follows.

Overview of theoretical approaches in 1963:

1. *Clinical approach* or *the theory of disharmony*: The main representative of this approach is the psychiatrist Lange-Eichbaum (first publication in 1928). In nine volumes (three more volumes will be published; this series is the seventh revised print!!) he demonstrated that psychopathology is an unavoidable correlate of giftedness. To be gifted means that you either already suffer from a psychological disorder or that you are in danger of becoming disturbed.

2. *The social psychological* or *socio-cultural approach*: There are two contrasting theories in this approach, namely the Theory of Acceleration and the "Zeitgeist"-theory. According to several authors (see Hill & Mönks, 1977), adolescence itself can be said to have been extended over the past decades by the earlier onset of the growth spurt as well as by earlier sexual maturation. It was assumed that this is accompanied by an increase in intellectual capabilities. In contrast, the "Zeitgeist"-theory assumes that there is a dramatic decrease in intellectual potential. Based

on empirical findings, the German sociologist K.V. Müller (1949) argued that students born between 1932 and 1937 showed a 2.1% decrease in intellectual potential per year. Figures for the 1932 cohort indicated that 17.8% were gifted; the figures for 1937 showed a percentage of 7.3%. In postwar Germany the view of Müller was supported for many years. Only recently has this view been modified as follows: there is not a decrease in, but a shift of cognitive abilities.

3. *The psychoanalytical theory*: According to this approach giftedness is seen as a "traumatically caused prematurity" resulting from not being able to solve the oedipal situation.

4. *The hereditary theory*: Based on some empirical findings, some authors concluded that there is a growing deficit in the reproductivity of the most gifted families. The conclusion was based on the fact that families with the highest intellectual potential have a declining number of descendants. Another conclusion, drawn from twin research, was that intellectual inheritance is relatively resistent to environmental influences.

5. *The theory of differentiation*: Related to the last perspective is the theory of differentiation. According to this approach the g-factor cannot be influenced by the environment to a great extent. However, the specific factors of intelligence (s-factors) can be shaped in a substantial way by the environment. In addition, cognitive abilities differentiate as age increases, i.e. after puberty the loading on the g-factor decreases and the loadings on the s-factors increase.

6. *The theory of harmony*: This theory of correlation between psychological and physical health goes back to the real pioneer in gifted research, Lewis M. Terman. In his dissertation (published in 1906) on "Genius and stupidity: A study of some of the intellectual processes of seven 'bright' and seven 'stupid' boys" Terman demonstrated his academic interest in giftedness. In 1920 he initiated a longitudinal project which examined the psycho-social development of 1,528 gifted children. This study is still continuing: it was taken over after Terman's death in 1956 by Robert Sears; and when he died in 1987 by Lee Cronbach. This is the longest and most thorough longitudinal study of the development of gifted people. At the beginning of this century the

prevalent idea about the "genius" was "early ripe, early rot". Terman and his associates provided empirical evidence that there is a harmonious development of gifted persons, psychologically as well as physically.

Mönks: Proposed longitudinal study, 1963

As a result of this literature review, which revealed such a diversity of (curious) opinions and theoretical analyses of giftedness, Mönks proposed the following longitudinal study with the intention to use the empirical findings for the improvement of psychological, educational as well as instructional guidance of the gifted:

1. Idiographic studies: Identification of developmental patterns in a life-span perspective.
2. Nomothetic studies: What are the developmental processes in gifted children? Are they different from normal children? What can we learn from precocious development for the understanding of 'normal' developmental processes?
3. What kind of interaction between nature and nurture occurs and what is favorable and what not?
4. What is the relationship between psychological and somatic development?

These were of course only general lines for a longitudinal study and would need elaboration. But, educational guidance and instructional programs for gifted are dependent on empirical findings coming from this kind of longitudinal research.

The article was like a cry in the desert. There was absolutely no public or scientific awareness that the development of the gifted, as well as the non-gifted, is to a great extent dependent on the facilities provided by the family and the school. It required another twenty years to make policy makers and the public aware of the special needs of the gifted.

1984: TOWARDS A STRUCTURAL APPROACH

The eighties witnessed new developments in the field of giftedness and the time was ripe for more permanent and structural solutions. The Dutch government supported several projects from 1984 onward (see also Span in this book). The very first project lasted from August 1984 until February 1986. The objectives were:

1) detection of gifted children in the first three classes of secondary education (age 12-15 years);
2) description of their social-emotional situation;
3) onset of qualitative differentiation.

Out of a representative sample, 772 students (394 girls and 378 boys) were selected. Twenty-two students (7 girls and 15 boys) were identified as gifted and 27 students (12 girls and 15 boys) were identified as gifted underachievers. (For a full report see Van Boxtel et al. 1986a and 1986b). In 1986, a pilot study was started which examined the gifted child in the elementary school (since 1985, entrance to elementary school is at the age of 4 in the Netherlands) (see Van Boxtel et al. 1987; Van Boxtel 1987). This was followed in 1988 with a project for curriculum development for the gifted in the elementary school. Up to the beginning of 1991 instructional programs were developed. They are now available at the Center for the Study of Giftedness at the University of Nijmegen.

It was also in 1984 that the first national conference on the gifted and talented took place in Nijmegen. Prior to this conference workshops and seminars were given by J. Renzulli and I. Sato. This first national conference - and since then the only one - brought teachers, scientists, policy makers and parents together. The interest shown by almost three hundred participants demonstrated that the topic of giftedness had been neglected and had remained outside public discussion for too long (see Mönks & Span, 1985).

Up to this point the scientific as well as the informational activities apparently had a central base in Nijmegen. It was becoming obvious that there was a tremendous public need for information and consultation. An advisory office was set up at the University of Nijmegen and two films on giftedness were made, entitled "Gifted - with ups and downs" (1988) and "Giftedness - research and education" (1988). Both films are available in Dutch, English and German.

1988: CENTER FOR THE STUDY OF GIFTEDNESS AT THE UNIVERSITY OF NIJMEGEN

The year 1988 would become an important one for the benefit of the gifted. Up to now, step by step progress had been slowly made. However, the academic and practical activities at the University of Nijmegen had now become so numerous that restructuring was required. The establishment of the Center for the Study of Giftedness in September 1988 was a logical culmination of the work done over the preceeding years. At the same time it meant that the end of one process was also the beginning of a new and ongoing development. The Center became a unit within the department of developmental psychology. The director of the Center was appointed to a newly established chair in "The development of the gifted child".

By giving the field of giftedness a clearly described identity within the university, the authorities of the university achieved something which was unique in Europe. This was the first time that the gifted child was regarded as a person who deserves and needs scientific and educational attention.

Now that the Center has been functioning for more than 3 years, an evaluation can be made which might be useful and important for the establishment of similar centers in Europe. The tasks of the Center are the following:

1. Research - Ph.D. and Master's theses form the main body of research. In addition, research is being designed to examine topics such as "Early development of gifted children", "Is the gifted child faster in his/her development and in what behavioral areas", "Emotional differences between the gifted and non-gifted child", "Self-concept development in gifted adolescents". Up to now the center mainly conducted applied research (see projects mentioned above).
2. Teaching - a course is offered in the graduate program on "Gifted children and adolescents - theory, research, practice".
3. Consultation/advisory tasks - since 1983 there has been a growing number of parents who need help for the education of their gifted child. Extensive diagnosis and advise is a daily task of the Center. The advisory task also includes consultation with teachers and schools.

4. Upgrading of teachers and professionals - since 1990 the Center has been offering a one day workshop for teachers, schoolteams and other professionals.
5 Documentation and information - the Center provides, on request, information about both scientific and practically oriented literature. A new magazine "VOORUIT" ("Ahead") was recently launched for teachers. This magazine appears four times a year and offers enrichment material for the elementary school for all grades and all subject matters. Related to this magazine, a newsletter is published, also four times a year, with background information about giftedness. Theoretical articles and news from the world of giftedness form the main content of this newsletter.

Finally, the Center was the main office for the organization of the Ninth World Conference on Gifted and Talented Children which was held in The Hague, July 29 - August 2, 1991. The Council of Europe chose the Center for the organization of the workshop "Gifted children and adolescents in Europe", of which this book is the result.

At the beginning of this chapter, some examples were given of the time lapse between the initial conception of some idea and its final implementation. The longest time lapse mentioned was 32 years (pace maker) and the shortest 6 years (video tape recorder). We also indicated that the first identified academic interest in giftedness in the Netherlands goes back to 1925 (Révész) and more specifically to 1938 (article by Waterink). The establishment of the first university center occurred 50 years later. We do not expect that it will require another 50 years before the Center for the Study of Giftedness has become a channel through which national and European organizations find their way to meet each others' needs. The result of the first workshop organized by the Center for the Study of Giftedness in cooperation with the Council of Europe will hopefully function as a catalyst, and thus give guidance and direction to those who commit themselves to the upbringing of healthy and happy gifted children, i.e. no repression or inhibition nor maximization, but appropriate guidance which facilitates optimal and healthy development for the individual.

References

Boxtel, H.W. van, Mönks, F.J., Roelofs, J.J.W. & Sanders, M.P.M. (1986a). *Onderzoeksresultaten, konklusies en aanbevelingen.* (Part I), Nijmegen: Center for the Study of Giftedness.

Boxtel, H.W. van, Mönks, F.J., Roelofs, J.J.W. & Sanders, M.P.M. (1986b). *Uitvoering van het onderzoek en verantwoording van het instrumentarium.* (Part II), Nijmegen: Center for the Study of Giftedness.

Boxtel, H.W. van (1987). *Hoogbegaafden in het basisonderwijs: een case-studie benadering in internationaal perspectief.* Nijmegen: Center for the Study of Giftedness.

Boxtel, H.W. van, Broekman, G.H.A. & Roelofs, J.J.W. (1987). *Hoogbegaafden in het basisonderwijs: een inventarisatie van de situatie.* Nijmegen: Center for the Study of Giftedness.

Duminy, P.A. (1959). *Begaafdheid - enkele psigologiese en pedagogiese aspekte.* (Giftedness - some psychological and pedagogical aspects.) Amsterdam: Jacob van Campen, (Dissertation Free University, December 1959)

Duminy, P.A. (1965). Begaafdheid en die onderwijs van begaafde kinders. Amsterdam: Meulenhoff

Hill, J.P. & Mönks, F.J. (1977). *Adolescence and Youth in Prospect.* Atlantic Highlands, N.J.: Humanities Press.

Lange-Eichbaum, W. & Kurth, W. (1985, Vol. 1;1989, Vol 7; 1991, Vol. 8; 1992, Vol. 9). *Genie, Irrsinn und Ruhm.* München: Ernst Reinhardt (7th revised edition).

Mönks, F.J. (1963). Beiträge zur Begabtenforschung im Kindes- und Jugendalter. *Archiv für die gesamte Psychologie, 115,* 362-382.
Bijdgragen tot de studie van de begaafdheid bij kinderen en jeugdigen. *GAWEIN, 12,* 76-103

Mönks, F.J. & Span, P. (eds.) (1985). *Hoogbegaafden in de samenleving.* Nijmegen: Dekker & van de Vegt.

Müller, K.V. (1949). *Die Begabung in der sozialen Wirklichkeit.* Göttingen: Hogrefe

Révész, G. (1925). *Psychology of a musical prodigy.* New York: Harcourt Brace. (Reprinted by Books for Libraries Press, Freeport, New York in 1970).

Révész, G. (1946). *Creative begaafdheid.* Den Haag: N.V. Servire.

Révész, G. (1952). *Talent und Genie.* Bern: A. Francke AG.

Terman, L.M. (1906). Genius and Stupidity: A study of some of the intellectual processes of seven 'bright' and seven 'stupid' boys. *Pedagogical Seminary, 13,* 307-373.

Waterink, J. (1938). Begaafdheid als kwalificatie der persoonlijkheid. *Algemeen Nederlandsch Tijdschrift voor Wijsbegeerte en Psychologie, 32,* 88-96.

Final Report

Herman W. van Boxtel[1]
The Netherlands

In this Workshop seven commissioned papers were presented in plenary sessions. Key issues, raised in these papers and their presentations, were then discussed in a number of sessions by three different working groups. In this final report some major points from the keynote papers are presented, followed by a summary of the discussions of the working groups. The contents of the keynote papers and the discussions are then integrated and conclusions drawn. Finally, several recommendations are made based on the results of this Workshop.

THE KEYNOTE PAPERS

In his contribution, **Pieter Span** concludes that in The Netherlands there is clear agreement that giftedness can only be identified after the production of outstanding achievement and only by specialists in the field. In the opinion of some researchers the concepts of "giftedness" and "expertness" coincide. This point of view makes it difficult to call children and early adolescents gifted since they have lacked the time to become an expert. Instead, the concept of "highly able" children is introduced. These children can be identified by their rapid cognitive development, their precocious knowledge base, and their outstanding metacognitive skills. These characteristics are seen as favourable for the development of gifted achievement, and are all open to testing.

An earlier model of giftedness, based on the ideas of A.D. de Groot, with giftedness heavily dependent on general intelligence (with creativity and motivation as inherent aspects, and seen as a natural potential largely due to inheritance) is no longer regarded as valid. Instead, researchers focus on other

[1] The author wishes to acknowledge the important and substantive contribution of Franz J. Mönks to the preparation of this report. The assistance provided by M.W. Katzko in editing the manuscript is also appreciated.

personality characteristics, such as creativity, divergent thinking skills, intrinsic motivation and task commitment, self confidence, as well as on other factors such as determination, persistance, and many, many hours of training and dedicated practice, as being necessary conditions for the emergence of gifted behavior. Very fruitful research has examined the behaviour of expert-novices, emphasizing metacognitive skills.

As a matter of fact, not viewing intelligence as a static and inherited personality trait opens the door for the possibility of stimulating and encouraging the development of gifted achievement by the environment. The important role of social settings, such as the family, the school, and the peergroup, is emphasized in the work of Mönks and his collaborators.

In practice, attention for giftedness began in the sixties, but was interrupted in the seventies. In the eighties, renewed attention resulted in a number of research projects. The findings of these projects suggest that both traditional intelligence tests and teacher nomination are unreliable identification instruments, that there is great need for enrichment materials and teacher training, and that gifted underachievers have very negative attitudes toward themselves and the school. At present, there is a clear preference for teaching the gifted in mixed-ability classes in the Netherlands. Unfortunately, teachers are inadequately trained and enrichment materials are insufficient. These are problems that should be tackled immediately. In the meanwhile, the best alternative is streaming in secondary education.

In contrast to Span's purely national focus, **Kurt Heller**, tries to give an overview of the international state of giftedness research. Nevertheless, his conclusions are remarkably similar to those of Span.

First, taking excellent performance in adulthood as a point of departure, Heller concludes that its individual and reliable long term prediction from childhood is extremely difficult, if not impossible. The main reason for this is that we try to predict area-specific performance on the basis of relatively general assessments of ability (i.e., IQ-tests most of the time). In fact, the relative significance of general intelligence potential for explaining performance variance decreases with progressing age, whereas elaborated area specific basic knowledge for explaining expertise becomes increasingly more powerful.

In attempting to determine and to explain the conditions which are favourable for the development of gifted behaviour, a central position is given to the analysis of the specific characteristics of the learning and performance behaviour of talented persons, and of the higher cognitive competencies (i.e. metacognition) that are assumed to be responsible for outstanding performance. Other parallels with Span can be found in the emphasis placed on divergent thinking processes, spontaneity and quick thinking (creativity), together with intrinsic achievement motivation and persistance. Moreover, these personality characteristics should be considered as the result of interactions between individual and sociocultural determinants, and which are to be examined in system-theoretical models of giftedness. Culture-comparative longitudinal studies deserve particular emphasis from this perspective.

In the eighties, a large number of basic and applied research projects were started/completed on (among others):

- the development of measurement instruments for identifying gifted children and adolescents;
- aid concepts inside and outside the school;
- social conditions and sex-specific differences;
- technical creativity and technological assessment;
- a variety of questions in cognitive psychology;
- cross-cultural studies in cooperation with China and the Russia.

Nevertheless, many topics have still been insufficiently investigated. In particular, the lack of basic research in the field of cognitive competence often makes applied research difficult.

In practice, support measures are directed predominantly towards enrichment. Acceleration approaches can be found in mathematics and occasionally in language, music, and sports. In the future, special attention has to be focused on:

- the development of talent-specific curricula;
- (advanced) training of teachers;
- training of school counselors, school psychologists, course and career advisors;
- the identification of and individual support for gifted children and adolescents.

Joan Freeman draws attention to the fact that in Britain a National Curriculum has for the first time been introduced. A possible risk of this policy is that the potential of children who could work beyond the expected attainment levels would not be fully developed.

With respect to education, there is no specific governmental policy for the gifted; nonetheless, interest appears to be growing. There is a preference for teaching the gifted in the regular classroom. The British school system has a number of characteristics which make it well suited for the gifted:

- flexibility in curriculum;
- thinking skills are trained by frequently writing essays;
- early school start: early reading and writing;
- efficient use of travelling specialist teachers (peripathetic teachers);
- there are many out-of-school extra's such as competitions, school orchestras and so forth;
- the school system is highly selective by implicit selection procedures in society;
- higher education is free.

Some very selective private schools could be viewed as schools for the intellectually gifted, although not originally founded with that objective. There is only one private school with a special curriculum for the gifted, and another one will be opened this year. On the other hand, there are several schools for special talents, such as music and performing arts. Acceleration, especially in the remaining grammar schools, is not uncommon.

Most research work on giftedness in the UK has been undertaken in the form of surveys. Some examples:

- teacher predictions for bright pupils;
- underachievement in gifted working-class boys;
- social influences on competence and achievement of the highly able;
- biographical determinants of musical excellence;
- Freeman's own well known longitudinal study (the Gulbenkian Project).

Provision for the highly able include:

- in education higher level thinking skills and effective problem solving are clearly promoted; there are workbooks for teachers and highly able pupils, especially in thinking skills;
- remediating highly able pupils with learning difficulties;
- the National Association for Gifted Children (NAGC), formed by parents, is active in running weekend educational enrichment sessions and summer schools;
- the National Association for Curriculum Enrichment & Extension (NACE), formed by teachers, has the aim of providing extra help for the highly able in normal schools;
- there is a number of private foundations and privately funded national competitions which are especially working for the gifted.

Andrzej Sękowski's contribution addresses the problems of gifted education in the countries of Middle East Europe in general and Poland in particular. The political system and ideology did not have a positive influence on the field of gifted research or on educational work with gifted children in post-war Middle East European countries. Unfavourable economic conditons and negative social attitudes are other factors which had a negative effect on interest in provisions for the gifted. At present, the author notes that attitudes are moving in a less negative direction.

Especially girls suffer from these negative conditions: in these countries sex seems to constitute an even greater barrier than it is in other European countries. Stereotyped judgement plays an important role. Although there are no limitations of an organizational nature which restrict the chances of girls for a successful professional career, the existing psychological barriers are often more difficult to overcome. Chances for girls are particularly unequal in mathematics and the natural sciences.

Basic research is primarily concerned with the nature of intellectual and creative gifts (creative and divergent thinking), cognitive style, and psychosocial conditions. There is some emphasis on the role of the sociocultural environment (especially the family), and on social functioning of the gifted. Sękowski calls special attention to the research on the gifted disabled; although it is valuable in its own right, it is fruitful for the gifted research field in general.

In Poland, gifted children are supported with scholarships from the Polish Children's Fund. They are allowed to follow individual programmes, to have contact with eminent scientists, and to participate in academic classes. However, they are not seperated from their own social environment, but are included in the regular educational schedule. The educational process is intensified through their participation in summer camps, where the objectives are the development of particular special gifts on the one hand and general psychological development on the other.

The education of gifted children in schools for the arts (music, fine arts) is very successful in Poland, as it is in most other Middle East European countries. Special program schools are also available for mathematics, physics, and language and literature. Special attention is given to the problems of gifted individuals from large rural areas.

The importance of East-West cooperation is stressed, both on a theoretical and practical level. On both levels a particularly useful contribution can be expected in identification and diagnosis.

Laurence Rieben discusses the situation in both France and Switzerland, a task made difficult by the heterogeneous situation (26 cantons in Switzerland with autonomous educational systems). An overview, partly based on a questionnaire, shows an almost complete absence of research in the field of giftedness in both countries.

The author defends the theoretical point of view that the development of gifted children can be explained in terms of general developmental models, provided that they include an explanation of individual differences. A crucial theoretical problem concerns the dimensionality of the models: do they have to be to considered as multi- or as unidimensional (g-factor)? The author prefers multidimensional models which define intelligence in terms of relatively independent components, which allows the possibility of several different profiles and developmental "routes". The question whether gifted children can be characterized by a prevalence of abnormal dissynchronicities in the development of these profiles (stressing the disharmonic or even pathological aspects of giftedness) is answered negatively. In the opinion of the author, much of this dissynchronicity, if present at all (in a higher degree than in the non-gifted population), can be attributed to the absence of adequate education adapted to the potential of gifted children.

Regarding the schooling of intellectually advanced children, the author discusses the fairly common practice in France and Switzerland of early entrance and skipping a school year after psycho-pedagogical examination, particularly in Geneva. This practice is considered as a suitable for most of the children passing the examination, but not really sufficient solution.

As a solution for the future, Rieben makes a plea for differentiated education. She mentions a number of prerequisites:

- development of an adequate differential psychology of education;
- development of know-how and technical means for identification of relevant individual differences within the classroom;
- encouragement of more flexibility in teaching approaches.

Such a solution appears to be taking form in France, allowing children to advance within a given course and within a class without having to skip a grade (the so-called "cycle system").

For ideological reasons there are no special educational procedures for specially talented school children in Sweden, at least since 1968. **Åke Edfeldt** discusses briefly the backgrounds of this -- as he sees it -- regretable situation.

The main part of his contribution is devoted to the immense importance of the parent's basic child-oriented efforts:

- most importantly during the early years;
- during the whole schooling period;
- by evoking intellectual interest and curiosity;
- learning the first lessons in basic problem solving.

He illustrates his point of view in terms of the process of learning to read. Quite a number of successful gifted individuals seem to have learned to read at a very young age, even before formal school entrance. He explains that learning to read naturally takes place in an analytic way. Helping the child in its first efforts requires no formal training and can be done by attentive parents and caretakers.

Analytically learning to read keeps the child free from the mechanisation of the synthetic school training in reading. It makes reading a true instrument for communication and thus for development as well.

In this sense, Edfeldt concludes that early learning to read (in an analytic way with the constructive help from parents and caretakers) can lead to academic prowess.

Candido Genovard reviews a number of research projects carried out in Spain since the beginning of the eighties[2]. As major findings of these projects the author mentions the following:

— the gifted subject (in homogeneous classes) is able to advance some twenty scholastic months during the nine months of the school year;
— prerequisites for a firm diagnosis of the gifted are objective measures of all the functions that determine or influence success in school;
— gifted students (partly from unfavourable home environments) seem to profit from the following (combined) provisions:
  - half-board,
  - homogeneous groups in primary schools,
  - quarterly supervision of performance by objective tests;
— the Renzulli rating scales for the behavioral characteristics of superior students appear to be efficient instruments for detecting students with high intellectual capacity.

In Spain the following organizations in the field of giftedness exist:

— the Research Team for Gifted Children (EINNS);
— the Association for the Development of Creativity and Talent (CREDEYTA).

Legislation allows gifted children to combine two school years in one or to skip a school year on the basis of an educational psychologist's report.

[2] A final version of this paper was not available for inclusion in this book.

Three groups, one English-speaking, one German-speaking, and one mixed group (using simultaneous translation facilities) were constituted at the start of the Workshop. In four working sessions, approximately eight hours of discussion were devoted to a number of topics. There had not been a list of topics and questions defined in advance to be examined by all the groups, although this did occur for some topics (such as the need for special teacher training).

The need for special attention for the gifted

Children differ in many aspects (for example rate of development, learning potential and learning speed, interests and motivation, special gifts and talents). Educational systems need to respect this extreme diversity and have the obligation to approach children in an individual way. It was noted that most countries in Europe have a fairly rigid school system, with school entrance and progress determined only by date of birth. Most educational systems do not treat each child as an individual and do not offer provision for the fullest possible development of each child's individual potential: "Educational legislation must recognise and respect the fact that children are different. The school system is to serve the children and not the other way around. If one of the two is to survive, it should be the child rather than the system".

Definition of giftedness

Little was said in the groups about what is in fact a gifted child. In the German-speaking group some alternatives were offered. One can define giftedness relative to the age group in question (the 2 or 3% highest scores on any identification instrument), one can take acquired expertness as a point of departure (problem: how to identify gifted children who did not have the time to become experts? Compare the contribution of Span), or one can apply subjective norms in a life span approach.

The lack of agreement on what constitutes giftedness poses a serious problem. One has to formulate a definition of giftedness before being able to identify the individuals who possess those qualites. Apparently, these difficulties were avoided by not explicitly discussing them, and/or the participants reached consensus on the subject they were talking about in a very implicit manner (which often is the case in discussions among experts).

However, there is one conclusion which can be drawn with certainty: giftedness is a multidimensional concept. In one group, a whole canon of aspects of giftedness was presented: cognitive abilities, creativity, communication abilities, learning strategies, social-moral abilities, self confidence, and the like. Although such a list does not clarify the definition problem, it makes clear that simple classical extreme IQ-definitions are no longer tenable.

Identification and diagnosis of giftedness

Without an explicit definition of giftedness, identification of gifted individuals is a difficult task. Even when we have a clear idea what it is that we want to identify, we still encounter many practical and methodological problems.

First, it should be noted that the identification strategies and procedures applied in the various countries of Europe show "enormous variety of methods and objectives". This variety seems to imply that the decision on what procedures and instruments to apply is dependent upon the existing situation in the educational system. This means that these procedures and instruments cannot be transferred without adequate knowledge of the conditions under which they are applied in other countries.

Second, until now much emphasis is put on the measurement of quantitative factors. This is even the case with characteristics such as creativity (for example in terms of the number of possible solutions offered for a problem), which might more appropriately be operationalized in terms of qualitative indices. There is a great need for the development of such qualitative indices.

Third, there seems to be agreement that all possible information should be gathered on all the relevant factors from as many different data sources as possible (parents, teachers, school psychologists, perhaps even peers) and by using a variety of instruments. In doing so, checklists and nomination procedures should be handled with discretion, as is the case with the administration of objective tests. Process and action oriented information is highly valued, in contrast to static status information. Identification of the gifted should be dynamic and continuous in nature.

Fourth, a number of relevant characteristics of the gifted can only be established when adequate learning situations are provided. This is especially true for such factors as creativity, originality, divergent thinking skills, interest, and motivation. The latter are hard to assess in standard testing situations and have to be evoked in regular classroom situations. Many teachers still need a lot of support to organize such learning situations. Furthermore, seemingly trivial information can prove to be of utmost importance, for example leisure time activity.

Fifth, the problem with many objective tests is how to fit them in regular school and classroom procedures, as well as the considerable cost of using them. Perhaps more emphasis should be placed on teacher training. Teachers should be made aware of the relevant characteristics of the gifted, thus making identification an ongoing classroom assessment and evaluation process, rather than a one-shot formal procedure.

Sixth, identification does not start at formal school entrance. As suggested in one of the groups "parents need to learn more about the early stages of child development in order to help them recognise signs of high potential". This does not mean that they have to be able to label their child as gifted or not. They have to learn to be sensitive and responsive in order to adequately foster learning and development (compare the contribution of Edfeldt).

Seventh, in identification procedures factors such as gender, belonging to ethnic and cultural minority groups, socio-economic background, and the possibility of learning disablities should be taken into account.

Finally, the need for more basic and applied research in the field of identification of the gifted is explicitly stated in all three groups. More collaboration between researchers and practitioners will be fruitful, not only in identifying the relevant questions to be addressed but also in putting research results into practice: "There is a need to bring together methodological know-how, pedagogical understanding, and practical support".

Provision for the gifted inside or outside the regular school system?

Having accepted the right of all individuals, including the gifted, to adequate educational provision (in the sense of fully developing his or her potential), how can adequate provision for the gifted be put into practice? In the working groups a wide variety of possible measures were discussed and proposed. These proposals ranged from out-of-school enrichment activities, enrichment in mixed-ability classes, in-depth study of subject matter, via acceleration and streaming to special curricula, and special schools and classes for the gifted. All these proposals "are considered more favourable than uniform teaching for all".

Yet, meeting the needs of the gifted within the regular school system seems to be preferred by the majority of the experts. Regarding the possibilities of teaching the gifted within the regular school system there is a further preference for teaching the gifted in mixed-ability classes. However, such a choice has important consequences.

First, a flexible curriculum is a necessary condition. However, in many countries in Europe curricula are fairly rigid, in fact too rigid to effectively match a great number of different teaching and learning styles (see also the contribution of Freeman on the characteristics of the British "open system").

Second, working with the gifted in mixed-ability classes assumes internal differentiation, which is in fact individualisation. To make individualisation work a number of suggestions were made:

- let pupils work at their own speed and rhythm;
- let gifted pupils make decisions in choosing learning materials and working methods;
- autonomous learning processes (independent learning) are considered very important;
- choose learning methods which stimulate motivation, create a stimulating environment;
- let classroom activities link up with leisure time activities.

Third, differentiation and individualisation means more than many teachers realize: "There needs to be a greater understanding about the variety of ways of differentiating work for pupils of different abilities". They also presuppose a great deal of practical knowledge and skill. Adequate teacher training seems indispensable (see below). Furthermore, team teaching and carefully managing the teaching and non-teaching staff is considered necessary in many instances.

Fourth, working with the gifted in mixed-ability classes can mean different things. One can choose to strenghten their strong points (preference model), one can choose to work on their weak points (remedial model), or one can choose to strenghten the strong points to compensate for weaker points (compensatory model).

Finally, it is recognised that working with the gifted in heterogeneous groups generally has a beneficial effect on the total group or class: "It can make an impact on a wide cohort of pupils...the standards are raised for all children".

There are a variety of other possibilities for working with the gifted within the regular school system. Alternatives discussed were acceleration, streaming, and special classes.

Acceleration is possible within a heterogeneous group of pupils (compare the French "cycle system", as described by Rieben). To accomplish the goal that pupils be maintained within the same group, some combination of acceleration and enrichment has to be organized. Good examples are the so-called pull-out models, allowing gifted children to work a certain amount of time outside the regular classroom.

Alternatively, acceleration can take the form of early school entrance, and/or grade skipping, a common practice in a number of countries. All the groups warned against the possible adversive effects of placing gifted pupils outside their normal age group. When such a form of acceleration is considered for a gifted pupil, a careful decision is needed, taking into account the whole social context, the willingness of the pupil, the opinion of the parents and the judgement of teachers and other experts. The same warning applies when placement in special classes is under consideration. Furthermore, teaching gifted pupils in isolation from their normal peer group may have negative consequences.

In many European countries streaming (placing pupils in more or less homogeneous groups) is often one of the few existing possibilities to meet the needs of the gifted, especially at the secondary level (compare for example the situation in Germany and the Netherlands; contributions of Heller and Span). It is noted that homogeneous groups often tend to become heterogeneous groups in the long run, so that solutions in terms of for example internal differentiation are as yet necessary. Another problem with streaming is that experts do not agree on the "ideal" moment for such a form of external differentiation.

Regarding the possibility of special classes, and particularly special schools for the gifted, most experts at the Workshop (as is the case for practitioners in the field of education) were very reluctant to adopt this alternative. The problem mainly concerns the social integration of gifted individuals. Teaching the gifted in isolation from their peer group could be harmful for their overall, and particularly social and emotional, development, if not properly managed. Furthermore, there exists a relation between optimal development of the gifted and the interests of society. In this regard, the gifted should be allowed and urged to accept social responsibilities.

A distinction has to be made between general academic giftedness (as it often is assessed by means of general tests of intelligence, exceptional academic achievement, or both), and special gifts and talents. Special schools for exceptionally talented children and youngsters in very specific talent domains (such as music and performing arts) are viewed positively in a large number of European countries. There is greater reluctance to adopt special schools for individuals with exceptional general academic capacities, particularly at the elementary level. This does not exclude the possiblity that in certain educational, political, geographical and economic circumstances such special schools may be appropriate.

The necessity of adequate teacher training

In the summary of discussions on teaching the gifted in mixed-ability classes, reference has already been made to the need for adequate teacher training. Teacher education and training was considered vital by all the working groups. It has the highest priority among the measures to be taken. So, a great deal of discussion was devoted to this subject, as was the case during the Vienna Summit of the World Council for Gifted and Talented Children in 1990, discussing the first stage of a strategic plan (see the Summit Report by Norah Maier)[3].

In teacher training, one has the choice between selecting and training a special group of teachers interested in working with the gifted on the one hand, and attention for the problems of gifted education in the training of all teachers on the other hand. The latter is clearly preferred in all the groups: all teachers will, as a matter of fact, have to work with gifted children, so differentiation strategies (enabling teaching the gifted in mixed-ability classes) should be an essential element of all initial teacher training. In most cases, initial teacher education was and is very limited, if not totally absent, so in-service training programmes are needed. It was stressed, that "such training needs to be carried out by those with both enthusiasm and expertise in the field of gifted education".

Important topics to be treated in both initial and in-service teacher training programmes that were mentioned:

* Awareness of the immense individual differences in, for example learning speed, depth of learning, and learning style, and the consequent needs of gifted children. Any negative images of giftedness, if already acquired, should be dispelled. Special attention should be paid to social stereotypes regarding the traditional role of women in relation to giftedness.
* Attitude change: "to meet the needs of the gifted, teachers must have an open mind, high self esteem, know their limits, be willing to learn together with the child, to favour independent learning, be motivated, and focus on the learning rather than the teaching process".
* Knowledge, insight and skills in a variety of strategies of differentiation, broadening the teachers' repertoire of choices for individualised educational care.
* Learning different teaching styles, and learning to effectively match different teaching styles and different learning styles.

3 Maier, N. (1991). *Vienna Summit Report: Quality and equality in the development of human potential.* Toronto: WCGT

Finally, all the groups emphasised that there should be closer cooperation between researchers and teachers. The complaint is often made that results of applied research are hard to put into practice, and that research projects in gifted education only have an influence in the institutions in which they are carried out. Results can and should be disseminated through in-service teacher training programmes.

INTEGRATION OF FINDINGS AND CONCLUSIONS

Theory and fundamental research

The number of existing models of giftedness is immense and continues to grow. From this Workshop one gets a good impression of the difficulties in defining giftedness. Yet it is possible to draw some conclusions:

- Capacity definitions of giftedness are out of date: conceptions of giftedness as being an extremely high score on some general test of intelligence are no longer seen as acceptable.
- Performance oriented models are gaining wide acceptance: outstanding performance has become the point of departure, sometimes equalizing "giftedness" and "expertness". For such performance, more is needed than extremely high convergent thinking skills traditionally measured by general intelligence tests.
- Research is focusing on the role of other intra-individual characteristics which can be seen as preconditions for outstanding performance, such as:
    * creativity and divergent thinking skills;
    * intrinsic motivation and task commitment;
    * self confidence and determination;
    * persistence, hard work, and dedicated practice.
- In no longer viewing intelligence as a static personality trait determined largely by inheritance, it is now possible to view gifted performance as a matter which can be influenced by the environment. The importance of social settings, such as the family/parents, school, and peer group, has increasingly been stressed in the last decade and throughout this Workshop.
- One has to avoid the danger that outstanding performance is too heavily interpreted in terms of academic performance alone. Other modalities

of giftedness (e.g. social, musical, artistic, psychomotor, linguistic) deserve equal attention, all the more because formal educational systems seem to be incapable of adequately handling these forms of giftedness.

Identification

Many problems concerning the identification of gifted children have been reported in this Workshop. To mention a few:

- Traditional intelligence tests are regarded as unreliable and of limited importance.
- In several research projects teacher nomination or teacher ratings also seemed unreliable.
- The assessment of creativity and divergent thinking skills poses many problems.
- On the one hand several questionnaires and checklists appear to be suitable; on the other hand they need to be handled with discretion.
- Only quantitative measures are available; there are very few qualitative approaches.

Generally, all available information with regard to all the relevant characteristics has to be gathered based on as many data sources as possible, using objective as well as subjective information.

Status information seems to be less important than process or action oriented diagnostics.

The question remains how formal identification strategies can be brought into the school and classroom. As one of the discussion groups pointed out, they might be out of date. Instead, to cite this group, "teachers need to be more reflective, adopting a technique of ongoing observation and using a repertoire of assessment and evaluation processes, in order to build up a profile of the child".

Provisions for the gifted inside the school system

With regard to the question whether nurturing the gifted should take place within the regular school system, or in special schools or classes, the predominant opinion in the Workshop appears to favour the regular school system. A variety of possible solutions were proposed. To mention the most important ones:

— internal differentiation,
— enrichment,
— more flexible arrangements and programs, so that regular contact with classmates is provided on the one hand, and activities and courses not related to age are possible on the other hand.

All these measures assume at least the following prerequisites:

— Adequate differentiated curricula and enrichment materials have to be provided. A great lack of these materials was repeatedly noted throughout the Workshop.
— Teachers have to be adequately trained.

The group discussions devoted much attention to a variety of problems related to teacher training. To sum up:

— Teachers have to be made more sensitive to individual differences in
  * learning rhythm and learning speed;
  * learning styles;
  * specific capacities and limitations;
  * specific needs and interests.
— Flexible teaching styles are needed.
— Teaching styles and individual learning styles have to be carefully matched; good relationships between teacher and learner are crucial in this respect.
— Many teachers have a limited view of what is meant by differentiation.

Differentiation has to be an important element of all initial teacher training. In most countries, this seems not to be the case. Additional in-service training programmes are thus essential. To be effective, such training needs to be carried out by those with both expertise and enthusiasm in the field of gifted education. Results of applied research can and should be disseminated through these in-service courses.

Acceleration in the form of early entrance, grade skipping or express-courses for homogeneous groups of gifted individuals appears to be common practice in a number of countries. These measures often seem to be taken because more adequate possibilities are lacking. Generally, teachers appear to be reluctant to use these measures. In all cases, possible negative social and emotional consequences of these measures need careful consideration. Maintaining contact with the peer group remains important, not in the last place in light of the social responsibility of the gifted. In making decisions for placement, all the involved persons (the gifted child, parents/caretakers, teachers, and other experts) should be allowed to take part.

Special schools and classes also appear to be acceptable in a number of countries. Here, one has to distinguish between special schools intended to meet the needs of students with exceptional general academic capacities (assessed by means of general tests of intelligence, exceptional academic achievement, or both), and special schools for exceptionally talented children in very specific talent domains.

The latter are acceptable in many European countries. That there are so few of these schools per country is related to the fact that these talents are regarded as "scarcity talents", and not because there is a principled unacceptability. Most often these special schools are for music and performing arts. Special schools for sports, mathematics, sciences, humanities (linguistics) are much less frequent, and if they do exist, it is predominantly in a number of Middle East European countries, and only at the secondary level.

General schools for the gifted are great exceptions to the rule in all countries of Europe. The general opinion on the necessity of these schools is consistent with their exceptional status: it is fairly negative. Nevertheless, in some countries there may exist political, geographical and educational circumstances which make special schools of this kind appropriate.

Provisions for gifted children outside the school system

Nurturing the gifted is not only the responsibility of the educational system. Responsibility has to be shared by the parents, the educational system, and other relevant institutions in society.

The parents have a very important role, especially in the first years of the child's life. They have to be sensitive to the needs of their children, gifted or not. They have to encourage and stimulate, and offer adequate opportunities for their gifted child. Education for parenthood in general, as an important aspect of secondary education, can be beneficial for the gifted.

As a group, parents of gifted children can do much to offer out-of-school provisions for the gifted. A good example is the British National Association for Gifted Children, which organizes weekend courses and summer camps.

In addition, parents of gifted children, including gifted problem children, need mutual support and recognition.

Some general considerations

The reality of individual differences is not always recognized and respected in legislation concerning the educational system. This seems particularly the case for gifted children.

Good education is education which is adapted to the specific needs and capacities of these children. The findings of basic research show that gifted children need a stimulating and encouraging environment to develop to their full potential.

Adequate educational opportunities are not only needed in the light of individual rights, but also in the light of the value of the gifted for society as a whole.

In addition, special attention is needed for the special and unfavourable position of gifted women.

We cannot afford to waste so much valuable talent.

RECOMMENDATIONS

1. Individual differences have to be recognized and respected in legislation. On the basis of individual human rights, gifted individuals need adequate educational opportunities to develop to their full potential.

2. The special and unfavourable position of gifted women and girls needs additional attention.

3. Provisions for the gifted and talented should preferably be arranged within the regular school system in the form of flexible curricula, internal differentiation, and enrichment activities.

4. There is a great need for the development of adequate enrichment activities, special curricula, and learning material for the gifted.

5. Adequate and effective teacher training is essential: differentiation in favour of the gifted should be a core element of all initial teacher training courses and supplementary programmes of in-service training in this area are badly needed.

6. Acceleration in the form of early entrance and grade skipping is acceptable in individual cases in the absence of more suitable measures, provided that it is arranged in agreement between school, parents, and child.

7. Special schools and/or classes for the gifted, in isolation from non-gifted children are -- as a rule -- only to be established for a number of special talents (e.g. music and performing arts). Under certain circumstances, such as for political, geographical or educational reasons, establishing special schools or classes of wider scope for the academically gifted or advanced may be appropriate and desirable.

8. Basic research in the field of giftedness has to be stimulated as a necessary basis for applied research.

9. Applied research is badly needed in the areas of identification strategies and provisions for the gifted; results should be made effective through initial and in-service teacher training.

10. Parents associations and other private initiatives in favour of the gifted should be encouraged.

APPENDIX

After the close of the workshop, some representatives of the Committee EUROTALENT expressed disagreement with the wording of some of the recommendations of the Final Report. They requested that their own modifications to these recommendations be included in the Report.

recommendation 3:

Provisions for gifted and talented should be arranged within the form of flexible curricula, internal differentiation, enrichment activities, acceleration and special classes.

recommendation 6:

Acceleration in the form of early entrance and grade skipping are recommended, provided that it is arranged in agreement between school, parents and child.

recommendation 7:

Under certain circumstances, such as for political, geographical or educational reasons, establishing special schools or classes of wider scope for the gifted may be appropriate and desirable.

PART 2
Commissioned Papers

Can early reading lead to academic prowess ?

Åke W. Edfeldt,
Sweden

In the introductory chapter to *Developing Talent In Young People,* Benjamin S. Bloom (1985) comments that "What any person in the world can learn, *almost* all persons can learn *if* provided with appropriate prior and current conditions of learning. This generalization does not appear to apply to the 2% or 3% of individuals who have severe emotional and physical difficulties that impair their learning. At the other extreme there are about 1% or 2% of individuals who appear to learn in such unusually capable ways that they *may* be exceptions to the theory".

Bloom also holds that the approximately 95% of school students who fall between these two extremes "become very similar in terms of their measured achievement, learning ability, rate of learning, and motivation for further learning when provided with favorable learning conditions", such as e.g. so called *mastery learning* in which students are helped to master each learning unit before they are allowed to proceed to more advanced learning tasks. Another even better set of favorable learning conditions might be brought about, if *tutoring* is used as the primary method of school instruction.

But are the favorable learning conditions of the most gifted 2 to 3% the same as the ones which the middle 95% of the student population require to succeed in their studies? In Sweden we just do not know. In Sweden it is considered undemocratic not only to be mentally gifted, but also to be gifted and demand special treatment because of this fact. And to be classified as undemocratic is in Sweden to be stigmatized with the horrible eighth amendment to the Seven Deadly Sins. L.M. Cohan (1985) observed that in Scandinavia there exist few provisions for the gifted child since "the egalitarian ethic is dominant and special provisions for the gifted are viewed as a means of maintaining the priviledged positions of certain groups". It is possible to trace this policy to 1968 and the work of a governmental committee chaired by a leading theorist of the Swedish social democratic party, Dr. Alva Myrdal, later Nobel Peace Prize winner. This committee was more interested in research and developmental work on the low achievers than on the talented children. Ergo, in Sweden there are, at least since 1968, no special educational procedures whatsoever for the specially talented school children.

But let us once more go back to Benjamin S. Bloom's interesting book, *Developing Talent In Young People,* in which he and his co-authors present their research on about how to create outstanding adult achievers out of highly talented children. Although this research must be understood within an American context, it depicts a general situation which is contradictory to the assumptions on which all earlier professional debate within this field has been based. As Bloom summarized (1985, p. 3): "The study has provided strong evidence that no matter what the initial characteristics (or gifts) of the individuals, unless there is a long and intensive process of encouragement, nurturance, education, and training, the individuals will not attain extreme levels of capability in these particular fields. This research has raised questions about earlier views of special gifts and innate aptitudes as necessary prerequisites of talent development."

If we turn our attention to the two academic subjects referred to in Bloom's report, mathematics and neurology, we find that the role of the parents was of immense importance during the entire school period of the gifted child, and especially so during the early years, but in a much more general and non-directive way than has been surmised before. The parents were *child-oriented*, meaning that they never hesitated to devote time and energy to their children's development. But when doing so, they were discerning enough to use an instructional format of play and pleasure. They were also aware that it was not their words but their actions which served as models for their children's behaviour. Of course, typical American middle-class virtues come into the picture too, like e.g., the parents' emphasizing the need "to *excel*, to *do one's best*, to *work hard*, and to *spend one's time constructively*" (Bloom, 1985, p. 510). The parents also had very strong interests in intellectual activities. This means that if they were not themselves actually involved in any of the learned professions, they at least spent a lot of their spare time engaging in some avocational intellectual interest. This might be seen as a sign of curiosity as a general attitude of the parents. And curiosity in the children seems to have been highly valued and encouraged by the parents. This is also what the parents of the fourty outstanding performers in mathematics and neurology seem to remember most -- their children's curiosity and innumerable questions.

The parents also possessed patience in answering their children's questions. If a question was too hard for a parent to answer properly, he or she made the effort to prepare an answer properly before answering the child or - even better - to have made it a family procedure to search for the correct answer in a book such as an encyclopedia. Therefore the children were not only given the proper answers to their many questions but they also learned the first lesson in all problem solving, i.e. always start by looking up what others have discovered before you. If that is enough, so much the better. If not, start testing your priorities or just start collecting enough empirical knowledge to choose your own way. This kind of problem solving as well as family discussions generally seem to be the most conspicuous features of these 40 academically gifted children's every day life.

In the very beginning they do not seem to have been exceptionally early to walk or speak. Some of them, maybe, but most of them were just normal and there were some rather slow to speak, too. The only comparatively common developmental feature seems to be that quite a number of these children learned to read at a rather young age. That not a majority did so obviously is the result of the schools outspoken claims for teaching reading as an exclusive privilege for trained teachers. Here are a couple of interview citations:

"He loved to have you read to him. I would set him in my lap and read. We had a lot of books for kids. I discovered that after I'd read a book to him a time or two, that I could stop in the middle of a sentence or in the middle of a paragraph and he'd go right along and finish it. He could remember it" (father of subject N-8).

"He would take a newspaper or book and write down words on a blackboard and ask me what they were. I can't take credit for getting him to do things, I didn't know whether or not it would be good for him. I just let him go on his own. I would read a story to him a couple of times and he would read it back to me" (mother of subject M-4). At four he was able to read and write.

My own interviews with the 1956 and 1957 freshmen of the faculties of arts and sciences of the Stockholm University, whom I have since followed up to 1982, show that the fifteen students with the best study records and the most prominent subsequent professional careers, all had told very similar stories about family discussions and joint problem solving as well as an early start in learning to read. Especially important they told of an analytical approach to their learning to read. This means that they learned to read in their eager pursuit of the content of a certain text or of the reason to another person's interest in a certain written material. These findings are also corroborated by analyses of the most proficient readers in my other longitudinal materials on reading as well as of the students with the highest ability to learn meaningful material which have long been studied by a group at the department of Education, University of Gothenburg, Sweden (Marton, 1975).

Again and again in scientific studies we find children, who learn how to learn without their own knowing it and many times even without their parents' knowing it, in their everyday conversation with their parents, in their joint problem solving and - not least in their analytic learning how to read at home and at a rather young age.

SCIENTIFIC BASES FOR THE ANALYTIC TEACHING READING MODEL

Passing reference has been made above to a specific theory of the reading process, called the *analytic* model. It is necessary to differentiate this theory from the competing *synthetic* theory which provides the theoretical basis for the conventional method of reading instruction using drill practice in phonics as its standard starting procedure. These two theoretical lines were both initiated towards the end of the nineteenth century, the synthetic one as a result of a British research tradition concerning reading disabilities and the analytic one as a result of the judicious reasoning which the American educational psychologist Edmund B. Huey (1908) performed with his starting point in the teachings of the experimental psychology pioneers, the German Wilhelm Wundt (1900) and the American William James (1892).

The method according to which most of us have learned to read, i.e. the synthetic, is thus characterized by its teaching phonics as a general starting point. At first the ability to decode the written signs is supposed to be taught, first one by one and later together in pairs or in small groups such as syllables. Only after the decoding is learned is the pupil supposed to take an interest in the content of the words.

This amounts to learning a totally new, artificial language, which later - and in the case of many children very laboriously so - is transferred to the understanding of texts in the reader's own mother toungue. During their first learning how to read the children are thus prevented from using the only existing, genuine support in the decoding task, viz. *the linguistic automaticity*, which through *structural learning* has been founded during the child's first years of life, i.e. when it was occupied fulltime in learning its mother tongue, or rather constructing its own, individual variation of the language spoken in the child's environment.

Whenever we speak, it is this linguistic automaticity which controls our pronunciation and prosody just as well as our lexical (semantic) and syntactic structuring of phrases, propositions and longer content periods. But the same linguistic automaticity also controls a listener's unconscious creative activity towards communicative understanding, i.e. the listener's interpretation. Furthermore it is the same linguistic automaticity which controls our unconscious choice of words and phrasings, whenever we write a short message or a more extensive text. And finally, it is this unconscious ability which renders it possible for us as readers to get the gist out of any written material in our mother tongue.

Now let us look at a very simplistic model of communication processes in general.

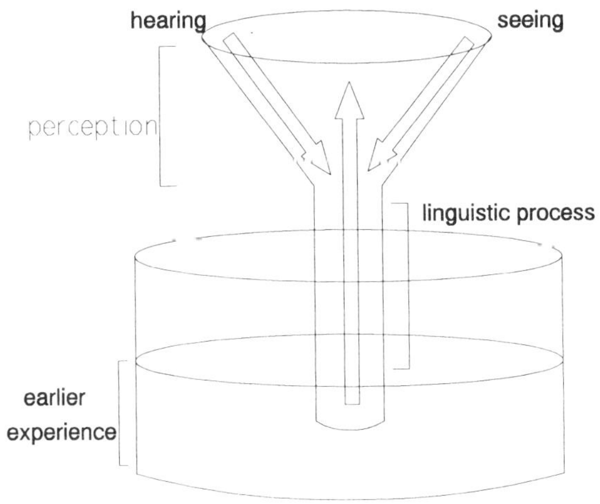

This figure depicts the individual as a funnel, with the wider top representing the perceptual phase of communication, the shank representing the individual's thought process, and the bowl representing earlier experiences in long-term memory. Research on the biological correlates of learning, recognizing, interpreting and understanding implies that although the incoming material takes part in the controlling activity during communicative processes, it is the accumulated experiences of the speaking, listening, writing or reading individual which is of the most significant importance in these processes. This figure thus illustrates a general law of all communicative activities: All communicative effects depend on the receiver's qualifications to interpret (=understand) and use (=act upon) the message, viz. on his or her accumulated earlier experiences.

This law is also basic to the analytic process model for reading and learning to read. In the analytic reading model a natural or preparatorily evoked preunderstanding exists *before* the decoding and thus controls the very process of decoding just as depicted in the funnel figure. But this means that you should never read or - even more so - engage in learning-to-read activities without a pre-understanding for the text to be read and a real interest in the content of this text. This implies that in analytic reading the communicative character of the reading process is always maintained. But how on earth do you manage to do that with the miserably watered down texts of an ordinary reader?

Such was my research project predicament when I happened to meet with representatives of the Stockholm group of daily newspapers. During our talks I realized that *the* solution to my problems was to use daily papers as reading texts instead of the ordinary readers and complementary textbooks on different subjects.

The basic difference between the two competing methods of teaching reading is thus the following. In synthetic methods of teaching reading "breaking the code" is the key ability. The written text is treated as if it were a new language, and thus the decoding procedure has to be overlearned before any questions about the meaningful content of the decoded text might be raised.

In contrast, analytic methods of teaching reading make optimal use of the automatic linguistic skills established during the process of acquiring one's mother tongue. This means that reading is treated as a communicative procedure, which in turn means that there exists a message to be decoded. The pupil's willingness to grasp this message, together with the mobilization of all possibly pertinent earlier knowledge (experiences) should always precede and thus control the decoding efforts.

As a consequence of this basic difference there are strictly speaking no demands whatsoever on texts to be used in synthetic reading. Of course it is nice if the texts are interesting, but since the childrens' decoding efforts always kill their genuine interest in any text content it does not really matter. In analytic reading on the other hand the necessary reading in depth, i.e., with the reader's controlling inner dialogue and preparatory text understanding running well ahead of and controlling the decoding procedure, it is a must that there are enough texts to choose from to allow every pupil to pick his or her own reading texts according to his or her own interests. This demand is not possible to meet with any plausible number of specifically produced school books. Here we thus find that only the continously produced daily newspapers can provide such a wide variety of texts to reasonable costs. For the preschool child any kind of printed material which catches the child's interest might be substituted for the newspapers, as long as for some reason - the child really wants to know what is printed there or what it is all about.

IS LEARNING TO READ A NORMAL ACTIVITY FOR THE PRESCHOOL CHILD?

Empirical evidence has convinced me that the systematic split which we see worldwide between mother tongue learning at home and learning how to read at school, i.e. between the lay sphere and the professional sphere, has negative consequences for the child's total linguistic development.

Elementary schools and the elementary schoolteacher profession were once concerned with teaching adults and, later on, children how to read. Eventually this specific task was passed on to an even more specialized professional group, viz. the female junior-school teachers. So definite was the teachers' claim to be the only ones capable of teaching children how to read that during the post-war period of 1950-1970, the Swedish State Board of Education as well as local school authorities warned parents of the dangers of trying to teach their own children how to read unless the parents had formal teacher training. Some local authorities bluntly forbade it. The reason given for this was that only trained professionals were capable of teaching children how to read in such manner that no specific reading problems (or so called word-blindness) would result.

Our most recent surveys on reading ability in university students seem to indicate, however, that the most able readers are most likely those who already knew how to read before starting first grade. The typical good reader is able to both consciously change his or her mode of reading according to the demands of the reading task as well as to unconsciously vary the mode of reading on the basis of text difficulty, its readability, as well as to other features of the reading environment, such as lighting or the amount of noise.

Through additional in depth analysis I was able to show that the good reader was continuously engaging in a sort of inner dialogue by means of which new reading material was being interpreted in terms of the reader's existing knowledge of the topic or, more generally, in terms of his or her accumulated experience. In actual reading this inner process also functioned as a preparatory understanding of the text material which followed the passage being read. In this preparatory sense understanding of the text was present before the decoding phase of reading had begun. As such the understanding served as a controlling factor in the decoding process.

Since really good readers carry on an "inner dialogue with the text", they are also able to silently read distinctly and slowly, which indeed diagnoses them as good readers. Almost all of them seem to have learned how to read while attempting to decipher messages from street signboards and milk packages or while sitting listening to their mothers. Thus they were able to escape the mechanizing effects of the phonetic drill method. But above all, their learning how to read developed as a natural continuation of their mother tongue acquisition. This would also seem to be the intuitively understood basis for several of the reading methods which appeared in the 1970's, e.g. the American LEA method by Hall (1976) and the Swedish LTG method by Ulrika Leimar (1974), which were at that time generally presented as producing "reading based on speech".

One could surmise that this total picture (early interest in letters and words as well as the qualitatively favourable outcome of parental intervention) is nothing other than an indication of giftedness. However, our experiments with entire schools in different socio-economic environments and, more important, in environments with different academic interests, have shown this not to be the case (cf. Edfeldt, 1989). These experiments were run as part of a three year longitudinal project starting with some 200 children from grades 1, 4, and 7. All reading instruction was according to the analytic process model. To emphasize that all reading has a communicative purpose the only reading materials used were daily newspapers from the Stockholm area. The controls -- around 7000 children

-- were from grades 1 to 9. The controls received conventional reading lessons, both for the beginners and for the older children's developmental reading.

In these experiments we found that the pupils who benefited most from our experimental reading instruction were the average and the less talented ones. The average gains which we found in experimental groups as compared to the control groups varied between 1.5 and 3 years of normal reading ability increase. This means, for example, that our experimental grade 1 pupils equalled the late spring status of control grade 5 by the time they had just reached grade 3.

The empirical contributions which finally made it possible to complete this rationale have been analysed at some length in Edfeldt (1990) in the concluding section entitled **Three Empirical Arguments.** If we accept increasing meta-linguistic awareness as a sign of linguistic development in general, we arrive at this very simple statement: children who learn how to read at school are generally taught synthetically, i.e. according to variants of the phonetic method. According to recent empirical studies at the University of Gothenburg, this kind of teaching succeeds in destroying within a year whatever linguistic awareness the children possessed before being taught to read by professionals. On the other hand, being taught analytically how to read in everyday life situations seems to stimulate the child's meta-linguistic awareness in two distinct steps:
a) the child becomes able to tell what talking means and what speech is used for,
b) the child becomes able to use this understanding as a starting point for explaining what reading means and what it is used for.

It takes no specific professional training to be able to answer the questions of a child who is attempting to understand the written messages which it stumbles upon in everyday life. An interest in the child and an understanding of what reading is used for is sufficient. Both these prerequisites are to be found in parents and preschool staff members. Thus, I cannot see any reason at all to object to a transfer of this educative task from the compulsory school to homes and preschools.

It is possible for even very young children to learn how to read, but such an achievement is of no avail if the child's general development does not allow it to make proper use of that reading ability. That is why we should never take the initiative in teaching a child how to read, but wait instead for the child's own signals that it wants help to facilitate its attempts to extract meaning from a written word or phrase. But then it is imperative that the child is given - with patience and care - all the support it asks for.

If this support is to be given by members of preschool staff, it must be remembered that for this task no formal training in teaching children how to read is required. The reform I am suggesting does not mean that the professional teaching of reading should only be initiated at an earlier age in the child's life. What it does mean is that it is time for the layman to enter a field of educational activity where the school professionals have been failing for far too long.

Just as the child learns to speak in its normal environment and at its own pace, with no other incentive than its natural urge to communicate, so too is the child likely to learn how to read. The only acceptable alternative is for the schools to replace the prevailing synthetic method with the recommended analytic method. But, as Edmund B. Huey (1908) stated almost ninety years ago, school authorities seem unwilling and all too slow to consider any alternatives at all. But here, it seems, might be one of many keys to the hidden "enormous potential pool of talent availabale in the U.S.", which was of concern to Bloom (1976).
L.M. Cohan's (1985) article on education of gifted individuals ends by saying that we have to address the early developmental steps which actually lead to academic superiority, rather than search for divine gifts or *the* educational procedure. In my opinion we find such a step - and a very important one too - in early learning to read, which in itself makes it natural for the individual child to carry it out in an analytic way. This means that from the very beginning the child commands a learner's instrument, which is kept free from the mechanization of the synthetic school training in reading, a true instrument for communication and thus for development too.

References

Bloom, B.S. (1976). *Human characteristics and school learning*. New York: McGraw-Hill.

Bloom, B.S. (ed.) (1985). *Developing Talent In Young People*. New York: Ballantine Books.

Cohan, L.M. (1985). Education of the gifted. In Husen, T. & Postlethwait, N.T. (eds.), *International Encyclopedia of Education*. Oxford: Pergamon Press.

Edfeldt, Åke W. (1989). *Educating young people with newspapers instead of books*. Institute of Educational Psychology, Stockholm, Sweden.

Edfeldt, Åke W. (1990) The Huey Legacy: A Cognitive Evaluation, *Early Child Development and Care, 59*, 53-72.

Hall, A.M. (1976). *Teaching Reading as a Language Experience*. Columbus, Ohio: Merrill.

Huey, E B (1908). *The Psychology and Pedagogy of Reading,* New York: MacMillan

James, W. (1890). *Principles of Psychology*. New York: Mac Millan

Leimar, U. (1974). *LTG* (Reading based on speech; RBS). Lund.

Marton, F. (1975). The Göteborg Project on Non-Verbatim Learning; *Report of the Institute of Educucation of the University of Gothenburg, 44*.

Wundt, W. (1911-1912). Völkerpsychologie I. Die Sprache; Leipzig: Engelmann.

Recent developments for the highly able in Britain

Joan Freeman
United Kingdom

The current great change in British education is the introduction of a National Curriculum, embodied in the 1988 Education Reform Act. Unlike other European countries, Britain has never previously had one. It became effective in the 1989 school year, guaranteeing more information to parents. It has a defined framework in which teachers still have their traditional flexibility to make many decisions on what and how to teach. However, the procedures by which the government's prescription is to be fulfilled are not yet defined particularly in terms of provision and assessment. Much has still to be done in the design of the National Curriculum: In fact, it has become an industry in itself, producing books and even university courses (Pring, 1989).

Parents, however, are wary of changes. In february 1991, a large group in Scotland, where the National Curriculum is not even in force, exercised their right by refusing to allow their children to be assessed at the age of seven. For their part, teachers are also objecting to the complexity of the assessment procedures, becoming exhausted, as they struggle to keep abreast of the Act's 238 clauses and 13 schedules, striving to understand the new rules, reorganise school administration, and still teach their pupils.

To add to the teachers' burdens, it is now possible for schools to remove themselves from Local Authority control to be funded by the central government. This obliges head-teachers to balance large budgets and to attract more children;

a few schools are even selling advertising space on corridor walls and the backs of exercise books. The outcome of these changes has been some return to more old-fashioned methods in education, i.e. towards more didactic teaching and less creative activity, which many schools believe is what parents really want.

There is some concern to know how the highly able are to be catered for in this new situation, and whether previous educational research findings based on the old system are still relevant, if children are now being taught different material by different methods. The National Curriculum does state that children should be educated to the level of which they are capable, though there is little specific guidance on how the needs of the brightest children are to be met. However, the new curriculum does have a potential to be of value to the highly able, through its emphasis on working across the curriculum. Its stated premise, though, that learning strategies involved in differentiated tasks and activities are transferrable from one to the other - is contentious. But the system is designed to be flexible, and so will be able to adapt to different circumstances.

The National Curriculum demands that every school-child must study the core subjects of mathematics, English, science, history, geography, music, art, and physical education, as well as a foreign language at secondary level. Each subject has ten defined levels of attainment to be assessed largely by teachers, which a child should be able to reach within certain age-spans. In theory, the needs of the individual are to be given the highest priority, each child working through the levels from its own starting point, at its own pace, and to the extent that it is able. This ought to be a truly differentiated approach, but it may restrict the ablest. For example, because the assessment of attainment is defined by a national age average at 7, 11, 14 and 16, it may concentrate teaching those prescribed standards, as though they were the limits of what might be expected at those ages. As a result, teachers may not realise that the attainments of some children could be well beyond them.

Interest in the gifted is growing slowly but steadily in Britain, especially since the 1989 Annual Report by Her Majesty's Inspectors of Schools (DES, 1989), which stated that able pupils were not being sufficiently stretched in most schools. In April 1990, the national Department of Education and Science (DES) brought toghether over 100 of the country's senior educationalists, specifically to discuss ways of adapting the National Curriculum for the highly able. This year, The City of Westminster Education Authority in London has produced a policy statement on the education of the highly able. The School Inspectorate is now visiting Local Education Authorities and individual schools around the country to investigate this aspect of educational provison. There is official recognition of a need for enrichment programmes for both slow and fast learners, and a shift of emphasis in teacher training and practice from the techniques of teaching to the ways in which pupils learn.

PROVISION IN SCHOOLS

National spending on state schools has dropped by nearly 8% in real terms since 1979. On the other hand, private schools have claimed an unprecedented increase of 1.7% of the school-age population in 1988 alone, so that they now enrol more than 7%. However, nearly a quarter of all these children have their fees subsidised - through their Local Education Authority, or the school's own scholarships, or through the national Assisted Places Scheme. This last is a major government input of about £57 million to provide for the educational needs of highly able children of modest means: its purpose is to remove such children from state education and pay for them to go to private schools. Naturally, teachers in state schools are less than pleased with this assumption that they are incapable of teaching bright children, or with the removal of their best pupils from the classroom (Whitty, 1989).

In Britain, there are now about 35 Local Education Authorities, out of a total 108, who employ at least one person with official responsibility for the highly able: some authorities have been active in this area of education for over a decade. In these areas, bright children can receive enrichment in many subjects; mostly offered out of school, in weekend and summer courses, either provided free by the Authority or privately at about 12 centres across the country.

There are schools for specifically talented youngsters in Britain, such as the Yehudi Menuhin and Chetham's schools for music, and several for the other performing arts. Although provision is also made for talented children in these subjects in normal schools, it is very variable in quality and quantity and often subject to the first cuts when local authorities are in financial difficulty. In theory, music tuition is freely available in British schools, in most there is a small charge for tuition from a peripatetic teacher and hiring an instrument, and in others it is all free. The standard of music in Britain is very high, although many schools do not take advantage of this provison, and facilities have suffered recently in some areas because of the problems of local government finance (Cleave & Dust, 1989).

Some very selective private schools could be said to be schools for the intellectually gifted, though they have neither claimed this description for themselves nor is their curriculum much different from that of other schools. However, when the whole school consists of youngsters within the top intellectual 2% of the population, the pace and depth of its teaching will clearly be different from that of a normal "all-comers" school. There is at present only one private boarding school in Britain which has a curriculum designed for the all-round education of gifted children - Cademuir International School, Peebles, which opened in September 1990. A day-school for high IQ scorers, Newton Preparatory School opened in London in September 1991.

Another new scheme is the City Technology Colleges, extremely well equipped secondary schools in urban areas, taking in eleven year-olds partly on motivation, and partly on parental commitment judged by interview. The colleges insist on a high standard of discipline and attendance, and encourage 16 year-olds to stay on at school by providing a wide choice of courses, including several of the colleges which focus on the non-science areas of moral values, the performing arts and the humanities. They were designed to be partly financed by industry, and are free to pupils. However, to date about 50 firms (one tobacco manufacurer) have pledged about £25 million, though some have either withdrawn their pledges or are not expected to pay. The government had to add £86 million, which does not come from the national education budget, to establish them and keep them running for the first 3 years. From then on, they are only to get about the same as ordinary schools for upkeep. The idea is to boost science and technology teaching, as well as offering close links with industry for work experience.

There are still 3.5% of all children in academic grammar schools (like the continental gymnasia) for which a Local Education Authority selects about 25% of the brightest children at eleven years-old (the "eleven-plus" exam). From 1946 this kind of selection covered the whole country, until comprehensive (all-comer) schools were created in large numbers in the 1960s. Although there were several reasons for this change, such as the unacceptable error rate in selection, many men and women are enjoying today the fruits of their grammar school education. Indeed, that pronouncement of pass or fail at 11 years-old still affects the self-concepts of millions of adults. The great majority of grammar schools used to be single-sex - a characteristic which has been shown by research to benefit girls' achievements, particularly in science. However, only 258 state secondary schools for girls remain in England and Wales today, out of a total of 4,250.

Comprehensive schools, though, do have some internal selection, dividing children for differentiated tuition. Some have "grammar"streams, others "vertical streaming", where children of different ages but at a similar stage of learning will come together for specific subjects. There are other variations, such as frequently used "setting", which brings groups of same-age children toghether who are at the same stage for tuition. Primary schools are not normally streamed, though the class teacher often employs a more subtle and flexible distribution of children into small groups, according to his/her estimate of their abilities.

Increasing the pace of children through the school curriculum, "acceleration" is not uncommon for gifted children in British schools. Individual academically children are most usually treated in this way, but some academically-orientated schools with a selective entrance system accelerate up to half their pupils for the state examinations, which are normally taken at 16 and 18. Other schools may have a signle fast stream, or even one teacher working in this way in his/her own subject, such as mathematics. Taken in groups, acceleration can be accomplished without the accompanying problems of social immaturity, poor peer relations, small physical size, etc., which often occur when one child is very much younger than the rest of the class. Acceleration for individual children in one or more subjects may also be offered when the school can accommodate this into its time-tabling.

It is strange that although Britain is somewhere near the bottom of the European league table for the proportion of 16-18 year-olds in full-time education, it ranks around the top in the proportion obtaining degrees, especially in sciences and engineering (Pearson, 1990). It seems that whatever the children's experiences are at school, a high proportion of the brightest do go on to complete their higher education. Nevertheless, great efforts are now being made to encourage more youngsters to stay on at school.

RECENT RESEARCH

Surveys

Most British research work on the gifted in the past ten years has been in the nature of surveys. In Oxfordshire, a survey covering 11 comprehensive schools, (Denton & Postlethwaite, 1985) investigated teachers' predictions for bright 13 year-olds in four subject areas. It found that these were reasonably accurate (as measured by the children's state examination results) about English, French and maths, but not about physics. Five of the schools then tried out the complex curriculum enrichment materials which the researchers developed for those subjects, as well as using their checklists to help predict high ability. However, they had little success. More promising work is under way there in a joint project between the North Wales (Clwyd) and Oxfordshire Education Authorities of in-service action research; this includes the provision of special education for highly able young children within the context of the National Curriculum.

Underachievement in gifted working-class boys was investigated by Tony Crocker at Wolverhampton Polytechnic (Crocker, 1987). His findings showed that when they come from stable homes and their parents considered that they were doing well, they did not show any signs of distress, unlike underachieving middle-class boys whose parents had expected better school performances. Crocker concluded that it is not always right to "rescue" such boys from this underachievment; he is to examine his sample again this year (1992), ten years after the original research began. He is also currently supervising two PhD studies, one on gifted drop-outs from university, and the other on the development of a teacher's checklist for underachieving gifted pupils.

Social influences were seen to have significant effects on the individual competence and achievement of 451 highly able young adults (16-23 years) in a 7-year research study by Tony Cassidy at Nene College, Northampton (Cassidy & Lynn, 1991). He concluded that the development of cognitive-behavioural styles of attribution, problem-solving, and achievment motivation were affected by the interaction between home background, type of school, IQ, and personality. To overcome academic vulnerability, both material and emotional factors had to be reinforced by the involvement of parents. George Ilsley (1989) investigated primary school teachers' conceptions of giftedness, and showed that although about half were aware of it, two-thirds felt a lack of confidence about its identification, and would like more help. They suggested that the easiest subjects for recognition were maths, foreign languages and music.

By far the most extensive British research has been the long-term Gulbenkian Research Project of gifted control and children which bagan in 1974 (Freeman, 1979; 1991). The target sample was 70 children, 5-15 years-old, whose parents identified them as gifted without testing. Each of them was matched for sex, age etc. with two control children from the same school class. The first control group was of identical measured ability, while the second was taken at random in that respect. All the 210 children were tested, and they, their families, and their head- and class-teachers were visited and interviewed. The first objective was to find out why - although they were of identical measured ability - it was only the target group and not the first control group who had been perceived as gifted. It appeared that parental identification of giftedness, whether the children were actually gifted or not, was based on their difficult behaviour, which was in fact usually due to family problems. Although over a quarter of the children scored IQ 150 or more, their emotional adjustment was in fact found to be quite independent of their intellectual ability. However, in some cases, scholastic progress was significantly inhibited by the lack of adequate learning materials and tuition, as well as by the family's cultural lifestyle and emotional stability.

Development of provision for the highly able

Activities for the promotion of higher-level thinking skills in adolescence have been developed over the last two years by Philip Adey and his team at the "Cognitive Acceleration through Science Project" at Kings College, London. They see the key to successful learning as the development of metacognition, i.e. understanding optimum personal reasoning patterns for particular problems (Adey, 1990). This work is complemented by the most recent research data on the British Ravens Progressive Matrices (Raven, 1990). Effective problem solving was found to be not only a cognitive activity, but is heavily dependent on affective and conative components: a high Ravens score does not necessarily correlate with a high verbal score or with high academic achievement (as Freeman, 1991, also found). John Raven suggests that to accommodate these data, a new psychometric paradigm is needed to cope with the variety of talents in a highly able individual, rather than a single factor model of ability. This would require, he says, a change of outlook in much of the educational provision now offered to "gifted" pupils.

Diane Montgomery at Middlesex Polytechnic is currently up-dating her work books for teachers and highly able pupils, especially in thinking skills. She has also been concerned for many years to help highly able pupils with learning difficulties in reading, spelling, and writing. When they are identified early, her system of "multisensory mouth training" has been found to improve the clarity of children's speech. Her remediation work with older children concentrates on psycho-linguistics and cognitive processes. She reports improved motivation for school-work, and hence life-chances, in the indivduals who sampled these techniques (Montgomery, 1990).

The National Association for Gifted Children (NAGC) was formed by parents, and now has about 4000 members in about 50 branches across the country. Their overall aim is the recognition of highly able children, and the encouragement of appropriate provison for them within normal schools. They run weekend educational enrichment sessions for both the children and their siblings, as well as summer schools in specific subjects. Most of their funds come from members, but they also receive sponsorship for conferences, and for the research they commission, such as a 1989 postal survey of the attitudes taken by Local Education Authorities towards gifted children. From that survey, in spite of the low response rate, it was concluded that understanding of the term "gifted" was confused. Also, because concern for the gifted was bracketed with all "Special Education", the handicapped were given proportionately very much more, so that provision for the needs of the gifted was spread too thinly.

The National Association for Curriculum Enrichment & Extension (NACE) was formed for and by teachers in 1982, to enable them to provide extra help for the highly able in normal schools. It produces paper-and-pencil enrichment material for the highly able, which can be used, for example, if these children finish early in class. The Lancaster Curriculum Development Centre, under the guidance of Ralph Callow, has also developed paper-and-pencil enrichment booklets for highly able primary school children. NAGC and NACE have now combined forces at one address in Nene College, Northamptonshire, under the banner of The National Centre for Able and Talented Children.

The Potential Trust is a private foundation for highly able children. It runs lively and satisfying summer camps for them, charging only a small fee, as do Mensa, the organisation for high IQ people, and Craigmuir School in Scotland. There are also many privately funded national competitions which effectively provide for the gifted such as Young Scientist of the Year, Young Engineer, the BBC Young Musician of the Year, and the Cadbury's National Exhibition of Children's Art, which includes poetry. Local Education Authorities and individual schools have also received financial help from industry for specific projects, such as competitions.

All the above associations are members of The European Council for High Ability (ECHA), which is very active in Britain, mostly in specialist teacher training. In 1991, it provided a week of training in teaching the highly able, and also two workshops to coordinate research on excellence in music and on "The Thinking Curriculum". In addition, it is launching an advanced teaching Diploma in The Education of the Highly Able, in cooperation with the College of Preceptors, while Sherborne school in Dorset is to provide the base both for the emergent ECHA European Youth Jazz Orchestra and the ECHA Schools Division.

Research overview

In Britain, there seems to be an endemic tendency to select and categorise people which is unlikely to be overthrown by any edict. Two major influences can still affect the development of high potential. The first is the international problem of gender, which slots boys and girls into respective areas of the curriculum, though these are not necessarily the most appropriate for their abilities. If children are nominated by either parents or teachers as gifted, without testing, this will normally result in two boys being presented for every girl. The second factor is the cumulative effect of social attitudes, which was found in the Freeman Follow-up study (1991) to be at times handicapping to the gifted.

Research shows the vital importance of both material and cultural provision in the development of children's abilities to an exceptionally high level. However, administrators and teachers in Britain are only just beginning to accept the idea that highly able children do need special educational provision, so that this is still very uneven and largely dependent on individual efforts and charitable organisations.

Educating the gifted as all-round people who are to make the most of their abilities, while retaining balance in their lives, is not dissimilar from the normal educational aims for all children. Their different needs lie essentially in the intensity of their learning abilities, especially in their greater facility for breadth and depth of learning. This implies differences of teaching style, as well as the need for competent, imaginative teachers, who should also be aware of their pupils' emotional development. The most able require a consistent challenge, and there are a myriad of ways in which it can be provided, using the resources which already exist.

The most valuable aspect of British education for the gifted is its flexibility. Although one cannot predict the precise effects of the National Curriculum, it is unlikely to turn British teachers into obedient civil servants. Britain is also the home of educational movements which have permanently changed international attitudes towards children and their education. For example, the open-plan construction of almost all new British primary schools means that, with the coordinated guidance of a team of teachers, children are able ro read in the quiet, carpeted reading corner, experiment with sand and water at the other end of the building, or work with a teacher on numbers in a different section. However, although this system is seen to work well with extravert childen, it is less productive with the introverts, who prefer a more structured environment (Bennett, 1976). There is a limit to the amount of child-centring that any system can accommodate successfully.

Despite growing interest from the government in the gifted children who attend our varied schools, there is still no specific overall educational policy for this minority in normal schools in Britain, where teachers prefer to keep gifted children in the normal classroom. In fact, teachers are generally wary of any form of "elitism", whether of wealth or ability, and are unhappy about categorising children into ability groups; this is often because they have serious doubts about the validity of tests of ability and what may be concluded from them. In many schools, there is also a deep commitment to mixed-ability teaching, which calls for special skills, even though teachers can find little training to help them cater for a wide collection of abilities in one classroom. They see children primarily as children and not as potential saviours of the nation. As a result, they prefer special provision for the gifted to be at practically no extra cost, such as acceleration or withdrawal for enrichment; this is because they are concerned that such special education for the gifted might take scarce resources away from other children. Yet *all* children, especially the culturally deprived, are very dependent on input from school.

CONCLUSIONS

Education for the gifted within the normal school environment is heavily dependent on the way the time and style of lessons are organised to cater for pupils' different abilities. To function comprehensively, catering for every child, requires flexibility in administrators' minds and a system which can change with circumstances. Here, for example, are two new and workable ideas for education the highly able which emerged from the April 1990 meeting organised by Her Majesty's Inspectorate:

* Open learning centres could be provided in school or community buildings, to be used either in or out of school hours, and with teachers and equipment readily available for all who want to use it, i.e. without selection. They should be well coordinated and integrated with school life. These centres would also provide teachers with specialist practice, both before graduation and on an in-service basis.
* Team teaching could move in a new direction, in which the team could take a theme and divide it into different subject areas, like the spokes of a wheel, with teachers working together, integrating their ideas and teaching through each level.

The National Curriculum has also recognised the value of flexibility in education, and has outlined possible approaches to teaching highly able children which normal schools could use. It suggest that schools provide:

- support from another institution such as a college
- support from peripatetic or advisory teachers
- individual schemes of work
- pupils who have reached the highest level in some foundation subjects before the age of 16 could be allowed to study other foundation subjects instead of continuing with the ones they already know
- arrangements for putting a child in an older teaching group - acceleration

There have been recent attacks on the old British (and European) assumption that education needs no further justification, that as a process of freeing and extending possibilities in the unique, individual child, it is inherently good. Without it, anyone would be intellectually and economically disadvantaged. But education in that old sense is being adapted with a greater emphasis on technical training - to provide "manpower" for a more technological society. Falling rolls in schools have narrowed the choice of options for children, although that is supposed to stop them spreading their energies too widely, and may give them a more balanced education in the end. But educational cut-backs have reduced teaching resources in many places, if not in technology. High unemployment, too, means that many of the brightest working-class children (those most likely to get the available jobs) sometimes see little point in staying on at school to take more exams which they regard as worthless, and are leaving school at the earliest opportunity for what work they can get. The universities report a falling off of applicants from the lower socio-economic levels. It cannot be assumed that gifted children will fulfil their potentials without special help.

Although in Britain, there are some specialist activites for the highly able, the most urgent areas for research are to make survey findings practicable. Although there is plenty of excellent teaching, no controlled expertimental work is being undertaken with different kinds of provision for such children, especially in mixed-ability settings. This is vital to improving provision. The following suggestions attempt to provide a basis for experiments, the results of which would be capable of bringing about positive changes in practice.

Firstly, administrative - organisational procedures within schools, which would allow comparisons to be made between schools with similar populations.
Secondly, curricular - instruction within and outside school hours, to see which kinds give the greatest help to the highly able.

There are many ways of improving teachers' ability to deal effectively with their most highle able pupils, either during first-level training, or as in-service courses. The education of the highly able can be improved without heavy expenditure. This can be accomplished using existing teacher's centres, raising awareness and stimulating ideas through information, and providing more resources, such as enrichment materials, where possible. There should also be more contact and coordination between different education authorities. Of course, there must also be agreed criteria by which the provision for such children might be assessed.

Britain can learn from countries which are seen to be successful in enabling highly able children to achieve at world standards in many fields. But we have something to teach others too, such as flexibility in education, concern for the individual child, the provision of free, high-quality universities, and non-school educational provision for all who want it - facilities which have enabled excellence to develop in people from all walks of life.

References

Adey, P. (1990). Pulling yourself up by your own thinking, European Council for High Ability (ECHA) conference, Budapest.

Bennet, N.S. (1976) *Teaching Styles and Pupil Progess* London: Open Books.

Cassidy, T. & Lynn, R. (1991). Achievment motivation, educational attainment, cycles of disadvantage, and social competence: some longitudinal dates. *British Journal of Educational Psychology, 61,* 1-12.

Cleave, S. & Dust, K. (1990). *Sound Start: the Schools' Instrumental Service*. Slough: NFER Nelson.

Crocker, T. (1987). Underachieving, gifted working-class boys: are they wrongly labelled underachieving?. *Educational Studies, 13*, March.

Denton, C. & Postlethwaite, K. (1985). *Able Children. Identifying them in the Classroom*. NFER Nelson, Staines, Middlesex.

DES, (1989). *Standards in Education 1988-1989,* The Annual Report of HM Senior Chief Inspector of Schools. London: Her Majesty's Stationery Office.

Freeman, J. (1979). *Gifted Children: Their Identification and Development in a Social Context*. Lancaster: MTP Press, and Baltimore: University Park Press.

Freeman, J. (1991). *Gifted Children Growing up*. London: Cassell.

Montgomery, D. (1990). Underfunctioning able and gifted pupils. European Council for High Ability (ECHA) conference, Budapest.

Pearson, R. (1990). *The European Labour Market Review: the Key Indicators*. London: Institute of Manpower Studies.

Pring, R. (1989). *The New Curriculum*. London: Cassell.

Raven, J. (1990). The Raven Progessive Matrices in gifted education. European Council for High Ability (ECHA) conference, Budapest.

Whitty, G. (1989). *The state and Private Education: An evaluation of the Assisted Places Scheme*. Basingstoke: Falmer Press.

Giftedness research and education of the gifted and talented in Germany

Kurt A. Heller

Germany

THE INTERNATIONAL STATE OF GIFTEDNESS RESEARCH

The current situation in the research on giftedness and talent is characterized by two contrasting paradigms: the psychometric paradigm on the one hand, and the cognitive science paradigm on the other. This is especially apparent with respect to basic research. Both paradigms, however, should not be viewed as competing but rather in terms of how they complement each other (cf. Carroll, 1983; Sternberg, 1985).

Waldmann & Weinert (1990) expect the cognitive science paradigm to generate important information, especially about the differential psychology of thinking, relevant to the formation of theories on giftedness and talent. But there is a two-sided danger in the extreme inter- and intra-individual variability in behavioural characteristics of gifted persons: unjustified theoretical generalisations as well as an over-production of models, e.g. for explaining the problem-solving behaviour of gifted persons.

The attempt to isolate task-invariant and area non-specific performance predictors, a task which has long dominated the research tradition, is made more difficult by the enormous number of empirically identified cognitive processes. More recent research results indicate a high degree of flexibility in (more advanced) problem-solving processes, where compensatory processes are often observed to operate among different cognitive determinants. The operation of invariant ability patterns which are independent of the level of giftedness thus seem to be unlikely. The idea that higher-order cognitive competences are responsible for outstanding performances seems more plausible. Thus, it would be logical to examine specific structural and procedural characteristics of the learning and performance behaviour of talented persons.

Similarly, it has been shown in more recent creativity research that divergent as well as convergent thought processes are necessary for creatively solving complex, difficult problems. In other words, the simultaneous use of both components — with sufficient area-specific basic knowledge — increases productivity. In addition to a high level of intelligence and flexible knowledge, spontaneity and quick thinking together with persistence and intrinsic achievement motivation favour creative problem-solving in advanced achievement situations. With the aid of the Chance Configuration Theory by Simonton (1988), this empirical finding can be explained. Through combination or re-combination of available elements of a problem area, cognitive variability is created, resulting in new configurations of those elements. In the experimental diagnosis of intelligence on the basis of item analysis (Klix, 1983) when examining area-specific, complex problem-solving processes or in the construction of new types of exercises of average difficulty (e.g., Sternberg, 1985; Dörner et al., 1983), such new models can be tested and further developed. A systematic combination of general and differential psychological research would be especially promising, according to Waldmann & Weinert (1990).

Within the context of developmental psychology, it is especially interesting to understand how more or less area-specific knowledge can be established apart from relatively general competences of ability. The learning processes involved can cover many years and are also subject to various internal (e.g., motivational) and external (e.g., socialisation and situation context) conditions. For this reason and also because of possible compensatory effects, long-term predictions of performance based on excellent performance in childhood is very difficult. In addition, the relative significance of general intelligence for explaining the performance variance decreases with progressing age, whereas elaborated area-specific basic knowledge for explaining expertise becomes increasingly more powerful. Reliable prediction of individual achievement is thus only possible to a limited degree in childhood and early adolescence (cf. Mönks et al., 1986; Trost, 1986; Heller & Hany, 1986; Heller, 1991).

Apart from a few exceptions (e.g., Tannenbaum, 1983), social factors have so far been neglected in the conceptualisation of giftedness and talent. This theoretical deficit is linked to the psychometric research tradition where test scores (e.g., IQ scores) are regarded as invariant measures over the life span relative to an age-reference group. Even the modern cognitive componental analyses are affected by this false assumption. The culture-comparative research of cognitive psychology, however, indicates that individual personality characteristics are always to be regarded as an interaction between individual and socio-cultural factors.

If a psychological description of this interaction process were attempted, social factors would have to be taken into account in addition to cognitive and noncognitive (motivational) components of the individual. It is these abilities which enable an individual to activate (cf. Mönks, 1985) his potentials in the right social settings (especially of the family, school and the group of friends). On the other hand, these abilities are influenced by the stimulation provided by social authority figures. To develop and empirically examine such system-theoretical models of giftedness would be an additional worthwhile task of the basic research in the field of cognitive abilities.

From a methodological perspective, culture-comparative longitudinal studies are potentially very valuable in the context of giftedness research. Such research would provide important knowledge about social and cultural influences on the development of cognitive abilities as well as about beneficial vs. unfavourable socialisation conditions. Culture-comparative studies are also essential for examining the generalizability of theories. To increase our knowledge about interindividual differences in the field of cognitive abilities, (quasi-) experimental studies would be required, e.g. to shed light on observed sex-specific differences in talent and performance in childhood, adolescence and adulthood.

Applied research differs from basic research primarily in its aims and less in its methods. In the context of giftedness and talent research, applied research, for example, concentrates above all on improving the practical possibilities for identifying and stimulating talented children and adolescents. Applied research is based on knowledge which has been more or less substantiated through basic research; however, such substantiation is often insufficient. As a result, applied research is often subjected to a "pressure to act" on the part of real life. This problem should be an important matter of concern and argues for closer co-operation between basic resesearch scientists and applied researchers.

In the analysis of individual learning and achievement abilities, preference should be given to process diagnostic instead of status diagnostic methods. In particular, multidimensional studies would be required. An improvement in our ability to predict future development and achievement depends on a) sufficiently reliable and valid indicators of giftedness, i.e., of individual cognitive abilities, b) the quality of the measurement of these indicators, and c) the prediction model, which should also take account of moderator variables, critical life events, and random influences on the predictor-criterion relation. Furthermore, it would be necessary to pay attention to nontrivial ability indicators (cf. Berg & Sternberg, 1985; Klix, 1983; Stevenson & Newman, 1986). Such knowledge would be essential for optimizing practical instruction and education measures, the development of ability-related aid programmes and also for interventions in individual cases.

RECENT RESEARCH ON GIFTEDNESS AND TALENT IN GERMANY AND CONSEQUENCES FOR MORE EFFICIENT AID

Since the beginning of the 80s, far more research projects have been planned and carried out in Germany on Giftedness and Talent. The Federal Ministry of Education and Science (BMBW) in Bonn has funded research on topics such as the development of measurement instruments for identifying gifted children and adolescents, the testing of domain-specific vs. non-specific aid concepts inside and outside school, the extension of promoting competitions, advisory services, and also on analysing individual developmental processes and their social conditions of the learning environment and of sex-specific differences in giftedness (especially in mathematics, natural sciences and technology). With the support of the BMBW, national and international conferences and symposia on giftedness and the talented were organized (cf. Cropley et al. 1986; Wagner 1990b; Wieczerkowski & Prado, 1990). Furthermore, the Third European Conference on High Ability (ECHA) will take place in Munich in 1992. In the areas of technical creativity and technology assessment, field studies as well as individual and educational psychological laboratory experiments are presently being funded by the Federal Ministry of Research and Technology (BMFT) and by the Founders' Association for German Science. Other more basic research projects deal with problem-solving strategies and cognitive abilities of the gifted, their metacognitive development, learning and memory strategies, knowledge acquisition, structure of expertise and similar questions of cognitive psychology, and also with the basic neuroscientific and philosophical principles of cognitive abilities. These projects are closely linked with the quasi-experimental intervention studies on changing unfavourable styles of causal attribution and motivation (e.g. with highly talented girls in the fields of natural sciences and technology), and on the general stimulation of metacognitive abilities, or even with neuroscientific studies, e.g. on "residual abilities" and their substitution in the field of cognitive achievement in the case of brain injuries. These and similar studies are funded mainly by the German Research Society and the VW Foundation. Finally, some cross-cultural studies are being carried out, such as the German/Chinese comparative study on the development of technical creativity (funded by the VW Foundation) and the study on building up technical knowledge as a condition for creative problem solving, involving the participation of scientists from the United States, Canada, Israel, Japan, and China (funded by the BMFT).

This brief review provides a more or less representative picture of current research on giftedness and talent in Germany conducted at university and research institutes (e.g. the Max Planck Institutes). Other projects in the field of applied research aim at developing and testing academic and extracurricular aid programmes, talent-specific counselling concepts, carrying out competitions (e.g. the Olympic Games in mathematics), summercamps or qualification measures for questions concerning talented individuals in the course of an in-service training program. Research projects which meet strict scientific standards are, however, relatively rare, e.g. Birx (1988), Hany & Bittner (1989), Engel (1990), Wieczerkowski & Jansen (1990).

The overall view so far should not hide the fact that despite an obvious increase in research activities over the last ten years, many topics could not be or were insufficiently dealt with scientifically. The following research topics seem important to me with regard to their potential value in supporting the talented in Germany (cf. Heller, 1990b):

development of instruction concepts of talent didactics;

— curricular development for special academic courses, special classes or even special high schools for certain acceleration groups and for enrichment courses on the extracurricular support of the gifted, including their evaluation (Scriven, 1967; cf. also Scriven, 1980, 1983; Brandtstädter, 1990; Thierau & Wottawa, 1990);

— the construction of process diagnostic identification instruments and the testing of successive decision strategies vs. cluster analytic procedures, multidimensional classification models;

- construction of area-specific counselling tests for talented adolescents;
- longitudinal studies of the gifted including analyses of the living environment over a whole life span;
- study of leisure-time activities of talented adolescents and their influence on personality development;
- career problems of talented girls and women, especially in the fields of mathematics, natural sciences and technology;
- analyses of metacognition, causal attribution (achievement motivation), control of actions, the self-concept of talented adolescents, including sex-specific differences and problems, e.g. regarding situation orientation vs. action orientation (Kuhl 1981, 1987), self-evaluation of heuristic abilities (Dörner et al. 1983);
- quasi-experimental intervention studies for counselling and supporting gifted children and adolescents.

Given appropriate concern for these topics, the main function of research on giftedness and talent should not be forgotten: society's support for the personality development of the gifted individual on the one hand, and the responsibility of the individual to society on the other. Personal responsibility develops from special abilities and abilities to appropriately contribute not only to one's own benefit, but also to the efficiency of the social network and its national economic basis. Differentiation and integration are thus complementary principles. Friction is inevitable when both are not adhered to. The opportunities of a united Germany and its integration into the European and non-European international community will only be successful if the mental resources of our young people can be successfully motivated. This humanistically and economically important aim is gaining world-wide acceptance, which brings us to to the next point.

DEFICITS IN THE IDENTIFICATION AND SUPPORT OF TALENTED CHILDREN AND ADOLESCENTS

Problems in the diagnosis of and support or counselling of the gifted have to be considered together. With the exception of high risk groups, e.g. disabled talented girls or children of minority groups, the difficulties of differentiation in instruction and education requires special attention. Despite a number of useful attempts over past few years to identify and support talented children and adolescents, there is lack of convincing evidence of the validity of many techniques, especially those which experts judge as innovative. Here, the requirement-oriented diagnosis and support concepts (cf. Heller, 1984) could be useful, especially when they are linked to recent results of applied problem-solving research and knowledge psychology. On the other hand, possibilities of psychometric identification, e.g., in the aptitude testing paradigm, have by no means been exhausted (cf., Trost, 1990). There is often a considerable discrepancy between the theoretical (multidimensional) reference model of giftedness and the practical (one-dimensional) identification and stimulation strategies (see, e.g., Rost, 1991; Hany & Heller, 1991; Mönks, 1991). Independent of such controversies, the following topics for improving identification and support can be regarded as urgent (Heller, 1990b):

- advisory aid for talent-orientated education, especially at pre-school and primary school age;
- advisory concepts for talented students, their parents and teachers;
- early identification and support of the talented disabled;
- identification and support of gifted girls (especially in the fields of mathematics, natural sciences and technology);
- identification and support of gifted underachievers;
- didactic possibilities of differential, i.e. talent-oriented, instruction regarding cognitive instruction/learning contents;
- academic stimulation of creativity, in general and also in specific areas;
- general vs. area-specific thought training programmes;
- work training programmes (e.g. for gifted underachievers).

If support of the gifted is widely seen as aid for individual development, then satisfactory results are to be expected if efforts are interdisciplinary in character. An important guiding role can be played by empirical psychology (e.g., in planning the curriculum and developing instruction concepts) and to psychology (especially developmental and educational psychology as well as differential and diagnostic psychology) without underestimating the value of the contributions of other sciences and humanities.

ORGANISATIONAL MEASURES FOR IMPROVING THE STRUCTURE OF THE SUPPORT FOR THE GIFTED IN GERMANY

It is striking that the need of special steps for supporting the talented has, as a rule, been recognized earlier in countries with established uniform school systems than in countries with traditionally structured secondary schools. This observation also applies to postwar Germany where certain groups of gifted adolescents (e.g., in mathematics or physics, and also in sports or music) were often more intensively supported in the former German Democratic Republic than in the Federal Republic of Germany. In particular, the Federal States of the Federal Republic governed by the Christian Democratic Union, the Christian Social Union and the Free Democratic Party, have increasingly committed themselves to supporting gifted students in secondary schools since the beginning of the 80s although they clearly favour the tripartite secondary school system. In contrast, the Federal States which are governed by the Social Democrats react rather reservedly as to the question of supporting the gifted in public schools. In this respect, public opinion in Germany regarding support for the gifted is still influenced by the educational policy, i.e., it is often more oriented to party policies than to relevant requirements. This is especially obvious where (with the exception of the need for special education) all students of the same age group attend the same type of school, e.g., the obligatory elementary school or the optional comprehensive school from the fifth grade onwards. Here, opposition is at its strongest, at least against external differentiating measures.

In the old as well as in the new Federal States of Germany, the support efforts based on the enrichment principle are predominant (e.g. the Baden-Württemberg extracurricular study groups for supporting especially gifted and interested secondary school students with now more than 3000 participants annually, the Bavarian plus courses, holiday or summercamps). However, support efforts based on the acceleration principle are much rarer; they can still be found in mathematics (e.g. in Berlin, Rostock, Hamburg, and Ulm), and also occasionally in languages or music and sports. Academic organisational acceleration measures are reflected in the recently re-discovered possibility of shortening the duration of studies at school and university (e.g. the Rhineland-Palatinate model with "fast-train classes", and other variants such as lift courses), the eight-year secondary school with special requirements (for gifted students) in Baden-Württemberg from 1991/92 onwards, as well as the programme for university graduates which has been funded jointly by the Federal and State Governments since 1990. In this case, too, the shortening of the duration of studies at university for especially gifted graduates apart from other special qualification requirements is emphasized. In the academic area, it was the private Christophorus-Schule (secondary school), with a boarding school, in Braunschweig which was the first school to gain nationwide recognition. A more detailed description of the situation of the support of the gifted in Germany can be found in Weinert & Wagner (1987) and Wagner (1990a, 1990b).

Over the past few years, re-discovered topics in support for the gifted have also included sex-specific differences of giftedness and performance, especially in the fields of mathematics, natural sciences and technology (Wieczerkowski & Prado, 1990; Beerman, Heller & Menacher, 1991). Co-education has again become a current topic, including proposals for quota rules to increase female interest in science and research. I am very sceptical that this and similar planning measures will succeed. From the perspective of cognitive psychology, this must be questioned as long as talented girls and women, when compared to comparatively talented males, more often show deficits in achievement motivation, control of actions and in other self-concept variables which are responsible for activating an individual's talent potentials. In other words: organisational measures for the structural improvement in a study and vocational situation, must accompany individual commitment and personal willingness to change unfavourable patterns of motivation and causal attribution. Otherwise there is the risk that such 'outside' aid, by not supporting self-esteem, involves more danger than benefit by provoking helplessness reactions detrimental to self-esteem. An efficiency control of the measures of supporting the gifted, including an evaluation of undesired side effects, thus seems to be just as important as the creative development of concepts and structure-improving measures are for increasing competence which benefits the personality development personality.

In contrast to external measures of talent differentiation, intra-academic and instructional measures of differentiation were rarely criticized in Germany. Despite this consensus of acceptance, the difficulty in realizing a differentiation concept appropriate for all levels of talent in a mixed (heterogeneous) abilities class should not be underestimated. Recent empirical studies seem to support the conclusion that the more capable students are supported to a lesser degree in heterogeneous groups than homogeneous groups (Treiber & Weinert, 1985; Helmke, 1988; Weinert et al., 1989, 1990). In general, the analysis of the learning environment of gifted children and adolescents is an important research task. For this reason, a study (Rost, 1989) has been funded for elementary schools by the Federal Minister of Education and Science since 1988. Important information has also been gained in the Munich longitudinal study on the development of giftedness (Heller, 1990a, 1991) as well as in the evaluation study - running since 1986 and intended until the year 2000 - on the Baden-Württemberg support of gifted students from secondary schools (cf. Hany & Bittner, 1989).

Finally, there is the task of psychological counselling of gifted children and adolescents as well as their parents and teachers. So far, the Federal Ministry of Education and Science has funded two model projects: An advisory centre in Hamburg, with an emphasis on diagnosis of individual giftedness and individual counselling, and an advisory centre in Munich which offers individual counselling as well as programmes for special courses especially for technically creative students. Moreover, corresponding advisory services are offered to parents seeking advice for their gifted children on the basis of private initiative (e.g., at the University of Tübingen) or by individual regional associations of the German Association for the Gifted Child (now for the whole of Germany); the work is done mostly on an informal basis and without sufficient institutional support. In contrast to the model projects in Hamburg and Munich, the latter initiatives are rarely set up, supervised or evaluated scientifically. It is thus extremely difficult to make an objective assessment of these advisory activities, especially as there are hardly any publications which describe them.

PERSPECTIVES OF AN ALL-GERMAN RESPONSIBILITY FOR THE FUTURE SUPPORT OF THE GIFTED

The conference "Research of Giftedness and Talent and Support of the Gifted in Germany 1980-1990-2000" (Wagner, 1990b), organized by the Society of Education and Giftedness in Bonn, attempted not only to give an overall view of the 80s but, for the first time, also made it possible for a larger number of colleagues from East Germany to exchange information with their colleagues from the West and to discuss future tasks in the united Germany. There was agreement that there should be closer co-operation in the future, in research as well as in the practical field of supporting the gifted. In both cases, however, considerable difficulties emerged and it will still take some time to overcome them.

The study on metacognition, which has been jointly conducted by psychologists of the University of Leipzig and the University of Munich (and funded by the VW Foundation) since 1990, could be an example of other German-German co-operation projects. This project examines the developmental conditions of metacognition and metamemory as well as the corresponding influences of the social learning environment. Regularly conducted workshops of both project teams have proved worthwhile. They do not only serve to co-ordinate the work in the project. An open exchange of information and experience makes it easier to solve the technical research problems which exist in the new Federal States. These problems concern financing the research, applicational procedures regarding financial support from third parties, which are of concern to everyone. In addition to structural improvement measures, new research projects should be planned and defended before experts from the German Research Society and other sponsors, e.g. the VW Foundation, the Federal Minister of Science and Technology, and the Federal Minister of Education and Science.

In view of the fact that, in the field of research of giftedness and talent, independent basic research was rarely possible and freely applied research was not at all possible in the former German Democratic Republic, especially the younger colleagues from East Germany are confronted with the need to be integrated into the international standards and research strategies as quickly as possible. This seems to be specifically a German task, and of course not only in the context of research on giftedness. The postgraduate project for the qualification of especially gifted scientists of the newer generation (graduate students studying for a doctorate and postgraduates), recently brought to life by the Federal and the 16 State Governments, can take over an important function also regarding all-German responsibility.

Similar possibilities for the research of the gifted should be explored in secondary schools. Apart from existing aid projects (especially for the financial support of the gifted) and different programmes (e.g. competitions organized by the State and Federal Governments in different disciplines, summer academies or summercamps) the main focus of attention for future tasks should lie in the development of talent-specific curricula and a broad educational and psychological advisory service. Finally, the primary structural development requirements include advanced training of teachers, counselling teachers, school psychologists, course and vocational advisors for questions concerning giftedness, and above all the recognition and individual support of gifted children and adolescents. Uniquely German problems arise in view of the necessary new orientation of the school system in the five new Federal States. For coping with these and other tasks, close and trusting co-operation of all those in the reunited Germany is required. Many problems, however, cannot be solved only on a national level. Especially with respect to research of the gifted, European and broader international cooperation is essential, both in theoretical development as well as the more practical considerations regarding utility.

References

Beerman, L., Heller, K. A. & Menacher, P. (1991). *Mathe - nichts für Mädchen? Begabung und Geschlecht am Beispiel von Mathematik, Naturwissenschaft und Technik.* Bern: Huber.

Berg, C. A. & Sternberg, R.J. (1985). Response to novelty: Continuity versus discontinuity in the developmental course of intelligence. In H.W. Reese (Ed.), *Advances in child development instruction* (pp. 381-414). Hillsdale, NJ: Erlbaum.

Birx, E. (1988). Mathematik und Begabung. Hamburg: Krämer.

Brandtstädter, J. (1990). Evaluationsforschung: Probleme der wissenschaftlichen Bewertung von Interventions- und Reformprojekten. *Zeitschrift für Pädagogische Psychologie, 4*, 215-227.

Carroll, J.B. (1983). Studying individual differences in cognitive abilities: Through and beyond factors analysis. In R.F. Dillon & R.R. Schmeck (Eds.), *Individual differences in cognition,* Vol. 1 (pp. 1-34). New York: Academic Press.

Cropley, A.j., Urban, K.K., Wagner, H. & Wieczerkowski, W. (Eds.). (1986). *Giftedness: A continuing worldwide challenge.* New York: Trillium Press.

Dörner, D., Kreuzig, H.W., Reither, F. & Stäudel, T. (1983). *Lohhausen. Vom Umgang mit Unbestimmtheit und Komplexität.* Bern: Huber.

Engel, W. (1990). Förderung mathematisch begabter Schüler außerhalb des Unterrichts und durch Spezialschulen in der DDR bis 1990. Wie geht es weiter? In H. Wagner (Ed.), *Begabungsforschung und Begabtenförderung in Deutschland 1980-1990-2000* (pp. 211-226). Bad Honnef: Bock.

Hany, E.A. & Bittner, R. (1989). Ergebnisse der wissenschaftlichen Begleitung zum Modellversuch "Arbeitsgemeinschaften für besonders befähigte Schüler" in Baden-Württemberg. Vierter Zwischenbericht. München: Universität, Fak. Psychologie.

Hany, E.A. & Heller, K.A. (1991). Gegenwärtiger Stand der Hochbegabungsforschung. *Zeitschrift für Entwicklungspsychologie und Pädagogische Psychologie, 13*, 241-249.

Heller, K.A. (Ed.). (1984). *Leistungsdiagnostik in der Schule.* 4. Aufl. Bern: Huber.

Heller, K.A. (1990a). Die Münchner Längsschnittstudie zur Hochbegabung und einige Folgeprojekte. In H. Wagner (Ed.), *Begabungsforschung und Begabtenförderung in Deutschland 1980-1990-2000* (pp. 34-45). Bad Honnef: Bock.

Heller, K.A. (1990b). Zukunftsperspektiven der Hochbegabungsforschung. In H. Wagner (Ed.), *Begabungsforschung und Begabungsförderung in Deutschland 1980-1990-2000* (pp. 276-284). Bad Honnef: Bock.

Heller, K.A. (1992) (Ed.). *Hochbegabung im Kindes- und Jugendalter*. Göttingen: Hogrefe.

Heller, K.A. (1991). The Nature and Development of Giftedness. *European Journal for High Ability, 2*, 174-188.

Heller, K.A. & Hany, E.A. (1986). Identification, Development and Achievement Analysis of Talented and Gifted Children in West Germany. In K.A. Heller & J.F. Feldhusen (Eds.), *Identifying and Nurturing the Gifted.* An International Perspective (pp. 67-82). Toronto, Lewiston/N.Y., Bern: Huber.

Helmke, A. (1988). Leistungssteigerung und Ausgleich von Leistungsunterschieden in Schulklassen: unvereinbare Ziele? *Zeitschrift für Entwicklungspsychologie und Pädagogische Psychologie, 10*, 45-76.

Klix, F. (1983). Begabungsforschung - ein neuer Weg in der kognitiven Intelligenzdiagnostik? *Zeitschrift für Psychologie, 191*, 360-387.

Kuhl, J. (1981). Motivational and functional helplessness: The moderating effect of state versus action orientation. *Journal of Personality and Social Psychology, 40*, 155-177.

Kuhl, J. (1987). Action control: The maintenance of motivational states. In F. Halisch & J. Kuhl (Eds.), *Motivation, intention, and volition* (p. 279-292). Berlin: Springer.

Mönks, F.J. (1985). Hoogbegaafden: een situatieschets. In F.J. Mönks & P. Span (Eds.), *Hoogbegaafden in de samenleving*. Nijmegen: Dekker & van de Vegt.

Mönks, F.J. (1991). Kann wissenschaftliche Argumentation auf Aktualität verzichten? *Zeitschrift für Entwicklungspsychologie und Pädagogische Psychologie, 13*, 232-240.

Mönks, F.J., van Boxtel, H.W., Roelofs, J.J.W. & Sanders, M.P.M. (1986). The Identification of Gifted Children in Secondary Education and a Description of Their Situation in Holland. In K.A. Heller & J.F. Feldhusen (Eds.), *Identifying and Nurturing the Gifted.* An International Perspective (pp. 39-65). Toronto, Lewiston/N.Y., Bern: Huber.

Rost, D.H. (Ed.). (1989). Lebensumweltanalyse besonders begabter Grundschulkinder. Forschungsbericht Nr. 2, Band I. Marburg: Universität, Fak. Psychologie.

Rost, D. (1991). Identifizierung von "Hochbegabung". *Zeitschrift für Entwicklungspsychologie und Pädagogische Psychologie, 13*, 197-231.

Scriven, M. (1967). The methodology of evaluation. In R.E. Stake (Ed.), *AERA monograph series on curriculum evaluation, 1*, 39-83. Chicago: Rand McNally.

Scriven, M.S. (1980). *The logic of evaluation.* Inverness, CA: Edgepress.

Scriven, M.S. (1983). Evaluation ideologies. In G.F. Madaus, M. Scriven & D.L. Stufflebeam (Eds.), *Evaluation models: Viewpoints on educational and human services evaluation* (p. 229-260). Boston, MA: Kluwer-Nijhoff.

Simonton, D.K. (1988a). Age and outstanding achivement: What do we know after a century of research. *Psychological Bulletin, 104,* 251-267.

Simonton, D.K. (1988b). *Scientific genius. A psychology of science.* Cambridge: Cambridge University Press.

Sternberg, R.J. (1985). *Beyond IQ: A triarchic theory of human intelligence.* Cambridge: Cambridge University Press.

Stevenson, H.W. & Newman, R.S. (1986). Long-term prediction of achievement and attitudes in mathematics and reading. *Child Development, 57,* 646-659.

Tannenbaum, A.J. (1983). *Gifted Children: Psychological and Educational Perspectives.* New York: MacMillan.

Thierau, H. & Wottawa, H. (1990). Evaluationsprojekte: Wissensbasis oder Entscheidungshilfe? *Zeitschrift für Pädagogische Psychologie, 4,* 229-240.

Treiber, B. & Weinert, F.E. (1985). *Gute Schulleistungen für alle? Psychologische Studien zu einer pädagogischen Hoffnung.* Münster: Aschendorff.

Trost, G. (1986). Identification of Highly Gifted Adolescents - Methods and Experiences. In K.A. Heller & J.F. Feldhusen (Eds.), *Identifying and Nurturing the Gifted.* An International Perspective (p. 83-91). Toronto, Lewiston/N.Y., Bern: Huber.

Wagner, H. (Ed.). (1990a). *Begabtenförderung in der Schule: Pädagogische Modelle in der Diskussion.* Bad Honnef: Bock.

Wagner, H. (Ed.). (1990b). *Begabungsforschung und Begabtenförderung in Deutschland 1980-1990-2000.* Bad Honnef: Bock.

Waldmann, M. & Weinert, F.E. (1990). *Intelligenz und Denken. Perspektiven der Hochbegabungsforschung.* Göttingen: Hogrefe.

Weinert, F.E. & Wagner, H. (Eds.). (1987). *Die Förderung Hochbegabter in der Bundesrepublik Deutschland: Probleme, Positionen, Perspektiven.* Bad Honnef: Bock.

Weinert, F.E., Helmke, A. & Schneider, W. (1989). Individual differences in learning performance and in school achievement: Plausible parallels and unexplained discrepancies. In H. Mandl, E. de Corte, N. Bennett & H.F. Friedrich (Eds.), *Learning and instruction* (p. 461-479). Oxford: Pergamon Press.

Weinert, F.E., Schrader, F.-W. & Helmke, A. (1990). Educational Expertise: Closing the Gap between Educational Research and Classroom Practice. *School Psychology International, 11,* 163-180.

Weinert, F.E., Schrader, F.-W. & Helmke, A. (1990). Unterrichtsexpertise - ein Konzept zur Verringerung der Kluft zwischen zwei theoretischen Paradigmen. In L.-M. Alisch, J. Baumert & K. Beck (Hrsg.), *Professionswissen und Professionalisierung* (Braunschweiger Studien zur Erziehungs- und Sozialarbeitswissenschaft: Bd. 28 (pp. 173-206). Empirische Pädagogik, Sonderband. Braunschweig: Copy-Center Colmsee.

Wieczerkowski, W. & Jansen, J. (1990). Mädchen und Mathematik: Geschlechtsunterschiede in Leistung und Wahlverhalten. In W. Wieczerkowski & T.M. Prado (Eds.), *Hochbegabte Mädchen* (p. 134-151). Bad Honnef: Bock.

Wieczerkowski, W. & Prado, T.M. (Eds.). (1990). *Hochbegabte Mädchen.* Bad Honnef: Bock.

Intellectually and educationally advanced children: A few theoretical and practical indicators of the situation in Switzerland and France

Laurence Rieben
Switzerland

PRELIMINARY REMARKS: AN "EXPERT LAYMAN" PERPLEXED BY THE NON-EXISTENCE OF THE FIELD TO BE STUDIED

I should mention straightaway that, contrary to usual practice, these preliminary remarks are as important as the rest of the paper, as they address two points which made the writing of this paper very difficult.

The first point concerns the ambiguity of my task. In theory, the task assigned to the rapporteurs of this workshop was to report on research in progress and on the situation of gifted children and adolescents in the education system of their own country. For Switzerland, it would have been necessary to refer to 26 cantons and half-cantons each with an autonomous education system. Lack of time, as well as the fact that I did not regard an exhaustive study to be of great value in the context of this report, meant that I did not opt for this approach. Furthermore, because the organisers were unable to appoint a French researcher to carry out the same task for France, I was asked to also look at the French situation. The result is an analysis that is limited in depth, and in particular, is of "variable geography" since sometimes it makes geopolitical distinctions (some facts concern France and not Switzerland and vice-versa), sometimes it makes linguistic distinctions (France and French-speaking Switzerland contrasted with German or Italian parts of Switzerland), and sometimes the facts apply only to a single Swiss canton, namely Geneva. Factual considerations and the interpretations made should therefore be regarded with caution as they rarely apply to an overall situation.

A second, more important, point concerns the almost total absence of theoretical or applied research on gifted children and adolescents in the regions and countries mentioned. This paucity, which initially I merely suspected, was confirmed by the meagre results of my efforts to find documentation and studies on this theme.

As far as France is concerned, I sent a short questionnaire to most, if not all, of the universities with a psychology and/or educational sciences department[1] in order to ascertain whether research on gifted children or adolescents was being conducted. Of the 23 questionnaires I sent, seven returned, with five confirming that the subject was not being researched. Of the two remaining replies, one came from Rennes University where A. Lieury (in press) is studying the link between memory and academic success by comparing gifted children with children obtaining poorer results in memory tests, and the other came from Caen University where J. Drévillon is looking into exceptional academic success of children from disadvantaged social backgrounds (no mention was made of a publication). The same questionnaire was also sent to the Institut National de Recherche Pédagogique in Paris, along with a request for a list, collated from available data bases, of French studies carried out in the past 20 years; this produced equally scant results. The only experiment indicated, of which I was already aware, was J.Ch. Terrassier's project for the establishment of a pilot school for advanced children commissioned by the Ministry of National Education (1986) and which led to the opening of such a school in Nice in 1987. The sparse findings of this search for documentation lead me to conclude that in the past few decades research in France has concentrated almost totally on children with difficulties rather than on advanced children[2].

[1] The questionnaire was sent to the following universities: Aix -Marseille I, Bordeaux II, Caen, Clermont-Ferrand II, Dijon, Franche-Comté, Grenoble II, Lille III, Lyon II, Metz, Montpellier III, Nancy II, Nice, Paris V, Paris VIII, Paris X, Pau, Picardie, Poitiers, Rennes II, Rouen, Strasbourg I, Toulouse Le Mirail

[2] It is important to note that it was only in the second half of this century that generalised statistical studies revealed, in France, the existence of slow and advanced performance at school. Given the massive extent to which pupils repeat academic years, it is not surprising that researchers directed their efforts first towards studying learning difficulties. In Switzerland, such statistics do not yet exist in all the cantons.

In Switzerland, the picture is no brighter for gifted children. To my knowledge, there is no research programme of significant scope on this subject being conducted in Swiss universities[3]. However, I did come across two theses in preparation. The first is being prepared by M. Stamm-Riesen, at the Pedagogical Institute of Zurich University, under the supervision of Prof. Fend, and deals with the support given to especially gifted children in compulsory schools in German-speaking Switzerland. This study will include the results of an examination of the legislation in force in the cantons concerned and of the analyses of interviews with teachers. The other thesis is being prepared by A. Aubert at Geneva University and supervised by prof. Mounoud. It concerns the development of intelligence and makes a comparative study of gifted children and normal children sorted according to chronological age and mental age. However, this is not a genuinely Swiss study since the experiment is taking place in the United States (Gifted Child Clinic, Prof. M. Lewis, University of New Brunswick). In reply to my request for an inventory of literature from 1960 to the present day, the Swiss Centre for the Co-ordination of Education Research (Aarau) and the Institut Roman de Recherche Pédagogique (Neuchâtel) confirmed the extreme scarcity of research in this field. U. Stednitz, a psychologist specialising in the counselling of gifted children in Zurich, recently published a report on the question of the development of creativity (1989). Secondly, U. Moser (1989) is monitoring an experiment, begun in 1986, in a school in Bern involving the "skipping" of a class. Lastly, I have carried out two longitudinal studies of children who skipped a school year, which I shall return to later in this report.

Thus, it certainly seems that in France and Switzerland in the past 20 years there has been a massive emphasis on the study of poor performance at school while on the continent of North America psychological and psycho-pedagogical research on gifted children and adolescents has been growing. Could it be that the study of academic success is taboo, and if so, why? To answer this question one would no doubt need to make a sociological analysis in connection with the various conceptions of the democratisation of education and the provision of equal training opportunities.

[3] It should be noted that H. Gruber, an American specialist on the study of creativity, known for his works on Darwin and Piaget, taught in Geneva between 1983 and 1989. To my knowledge, he did not however carry out research into gifted children.

My third and last remark concerns the limitations of my expertise in the field (it is hardly surprising that a Francophone expert cannot be found if the field of research itself is inexistent!). Some of my contributions on the theme of the workshop date back many years and are relatively marginal since, for the past 10 or so years, my research has concentrated primarily on the study of individual differences in cognitive development, and more recently on the study of both the developmental and differential acquisition of written language by so-called normal children. I agreed to draft this report because the organisers persuaded me that my refusal would have meant that there would have been no Francophone point of view. I therefore came round to the idea that to establish that there was an absence of such research was better than to establish nothing at all and that it could serve to prompt research in the future.

My first contact with studies on giftedness[4], mostly American, date back 20 or so years, to a time when it was very common to associate giftedness with high IQ. My point of view at the time, which remains unchanged on this matter, questioned the notion of global intelligence and consequently to consider IQ as an unsuitable method of evaluating all the facets of cognitive development, in particular operational development and creativity, both in normal children and in backward or advanced children. There was nothing particularly original about this standpoint at the time in Geneva since it reflected a critical attitude towards the intelligence scales adopted by Rey (1963) as well as by followers of Piaget (Inhelder, 1963). I then became interested in studying the link between creativity and intelligence in children (Rieben, 1978; Rieben & Mengal, 1977). My study consisted of examining two groups of children (6 year-olds and 8 year-olds) using the WISC, two Piagetian logic tests and the Torrance creativity battery. The results confirmed the assumption of a relative autonomy between these three facets of cognitive development, which was greater at 8 years old than at 6 years old.

However, my interest in this problem and more particularly in the study of creativity, began elsewhere. Until 1989 I used to be responsible for the

[4] In French, it is difficult to find satifactory vocabulary in this field. First of all, the words "don" (gift) or "doué" (gifted) encourage a hereditist interpretation of development, which is not necessarily that of the psychologists interested in this field. Secondly, for want of a corresponding noun, one either writes "giftedness" in a French text or invents a new word. The French-speaking Canadians (see Gagné, 1987) opted for the second solution and proposed the word "douance" which I have adopted in the French text in the interests of terminological consistency.

psychological examinations for "age requirement exemptions", which were requested by the Department for State Education of the canton of Geneva when parents asked that their children be able to skip a school year. Between 1966 and 1989, several hundreds of children were examined in this way. My problem for many years, although I never had the material resources to study it systematically, concerned the content and the validity of such an examination. A first validation study was done in the 1970s (Rieben, 1980), and the second is still in progress; some of the results will be presented below.

It is therefore clear that my approach to the problem of gifted children is very limited. It is essentially confined to the problem of the examination of cognitive functions in connection with the question of age requirement exemptions. This does not in any way mean that I consider that congitive problems are the only aspects involved in the study of giftedness. On the contrary, in this field, as in others, it is necessary to integrate cognitive and psycho-dynamic approaches (see, e.g., Rieben, 1982).

The subsequent sections of this report will be set out as follows. I shall first present a few questions of basic research which have not yet been satisfactorily answered by the North American studies and which should therefore be the subject of future research. I shall then go on to more applied questions by presenting a few facts on the education of intellectually advanced children in France, Switzerland and in particular in Geneva. Lastly, I shall conclude by discussing the more general problem of differentiated education.

A FEW BASIC QUESTIONS ON GIFTED CHILDREN

Historically, the field of research on giftedness is related to that of differential psychology, i.e., to the problem of measuring intelligence and to the respective parts played by heredity and environment on psychological behaviour (Terman, 1925). As a result, few theories on giftedness emerged in the first half of the century. From the 1950s, Guilford (1950; 1967) contributed to the study of creativity. However, it is only recently that attempts have been made to treat the question in the context of general models of development (see for example Horowitz & O'Brien, 1985; Feldman, 1982; Freeman, 1985). Sternberg (1985), in considering this literature, shows that two perspectives are possible: that of the designing of models to explain specifically the development of gifted subjects or that of the extension of existing theories (for example by defining a post-formal phase in the context of the Piaget theory). For my part, it seems that a third perspective, corresponding to another type of extension, could consist of integrating the study of individual differences (in particular with regard to gifted subjects) into the general model of development. It should be noted that these models generally consider individual variation as background noise. In research on the cognitive development of so-called normal children conducted with S. de Ribaupierre (in Geneva) and J. Lautrey (Paris V) in the context of the Piaget theory, we considered, like Reuchlin (1969), that any general model must be able to take account of individual differences and that consequently the study of these differences contributes to the construction of general theories (Lautrey, de Ribaupierre & Rieben, 1981; 1987). With the same aim in mind, we studied adolescents performing poorly at school (Rieben, de Ribaupierre & Lautrey, 1985) but it would have been possible and probably valuable to study children who were ahead in their development. The idea that the study of gifted children, and particularly of the processes involved in their development and not only of their performances, can contribute to a better understanding of the development of intelligence in general is also taken up by Aubert (thesis in preparation, mentioned above), which deals with the processes involved in various classifications, problem-solving and perspective-taking tasks.

Connected with this first question, another more general theoretical problem which is crucial in defining advanced children concerns the unidimensional aspect versus the multidimensional aspect of development. In unidimensional models or those with a strong emphasis on a single dimension of development, advanced cognitive development can be defined by a single factor. Thus in a model emphasising the g-factor, a high IQ or an excellent result in a test heavily permeated by the g-factor would be sufficient to define a child as being gifted. By the same token, in a classic Piagetian model, it would be possible to emphasise an advanced operational level.

However, if intelligence is defined by a set of components revealing a certain degree of independence, one can assume that there are different developmental "routes" and present the problem in terms of cognitive profile. In such an approach, the definition of the advanced child will be more complex for at least two reasons. The first concerns the choice and number of components regarded as relevant, both for the purposes of the theoretical validation of giftedness or intellectual precocity as well as for the purposes of empirical validation; there is still much research to be done at this level. The second reason is connected with the very nature of the notion of profile which implies by definition the existence of possible discrepancies between different performances or different processes involved in development.

Although the line between "normal" or "subnormal" heterogeneity may not be sufficiently clear, heterogeneity as such has served to define specific groups of subjects, and the "décalage"[5] regarded as a symptom. This approach has been used to identify both cognitive shortcomings or backwardness as well as precocious development. In the first case, an example is the notion of heterochronism used by Zazzo (1965) in connection with mental deficiency or the notion of cognitive disharmony proposed more recently by Gibello (1976). In the second case, namely the study of advanced children, definitions in terms of individual variation outside the norm are very frequent. These can be found in the few French studies which aim to inform the general public by referring mainly to basic American works. Despite significantly different perspectives, the authors all stress the "disharmonic" and even "pathological" aspects of giftedness. Chauvin (1975) refers to "prodiges inquiétants", Terrassier (1981, 2nd edition 1989) writes about the "syndrome de dyssynchronie" and Coriat (1987) devotes a chapter to "surdoués inadaptés et handicapés". Without wishing to deny that such cases exist, the results we obtained in our research on cognitive development, prompted us to regard disharmony as the norm. If we look closely enough, considerable intra-individual variability can be observed among most so-called normal children (Rieben, de Ribaupierre & Lautrey, 1985). This variability is all the more interesting insofar as it can be interpreted theoretically; in the research mentioned above we sought to place it in relation to different modes of processing information (Rieben, de Ribaupierre & Lautrey, 1990).

Thus if we adopt the established belief that 5% of children can regarded as gifted (in other words approximately 200,000 in France for the five levels of elementary education and roughly 35,000 in Switzwerland for the nine levels of compulsory education), it would have to be ascertained that they were all more dissynchronous than children situated in the norm. I believe that the only

[5] Editor's comment: Décalage is a Piagetian term and translated from French, roughly, as discontinuity or slippage. The term is used for all forms of nonsynchronous appearance of concepts. For example, when a child shows conservation of number but not of mass. This kind of discontinuous cognitive development is called horizontal décalage.

problem for most of them lies in the fact that they do not receive education adapted to their potential. This point takes us to the next section which deals with the schooling of advanced children and, in particular, with being ahead in one's education.

THE SCHOOLING OF INTELLECTUALLY ADVANCED CHILDREN

Age and schooling

In most education systems, the chronological age of the pupils is the chief regulatory factor; however, the age of entry into equivalent school levels can vary considerably from one system to another, illustrating the fact that the rules are based on education policy choices rather than for scientific reasons. For example, it is interesting to compare the different ages at which schooling becomes compulsory (although most children begin school before the compulsory age), when the first basic skills, and in particular reading, are acquired (Cours Préparatoire in France, and 1ère Primaire in Switzerland, henceforth referred to as CP and 1P in the text).

Table 1 shows the proportion of children according to age[6] (including children attending private schools) in CP for France and 1P in Swiss cantons, first in 1980-81 then in 1989-90. As can be clearly seen, there can be a difference of a whole year between different school systems, which around the age of six-seven years old is not insignificant. In some places, for example Aargau, the majority of children are seven years old in 1P; in Ticino, virtually all the children are six years old. Neuchâtel is in an intermediate position as is Geneva. In France, the age proportions are more in line with those of Ticino.

Moreover, these figures show that there has been little change in the last 10 years and that the very slight fluctuations sometimes reflect an increasingly young population (in France or in Aargau) or an ageing of the population

[6] For France, these data have been taken from a document published annually by the Ministry of National Education: Repères et références statistiques sur les enseignements et la formation, année scolaire et universitaire 1988-89. Paris, Imprimerie Nationale, 1990. For Switzerland, the information was sent to me by the Federal Statistical Office in Bern. As far as the choice of Swiss cantons is concerned, this was necessarily arbitrary. I simply sought to highlight differences while taking care to select one canton from each linguistic region and adding Geneva which is analysed in more detail subsequently. Lastly, in both countries, ages are calculated solely on the basis of date of birth, thus a child born in 1983 will be described as being six years old during the academic year 1989/90.

(Ticino or Neuchâtel). Lastly the very low proportion of five year-olds (in France and in Geneva) show that these education systems practise "age requirement exemption" whereby a child may enter a school year early. This is the next subject I shall deal with.

Table 1

Percentages of children by age
CP (Metropolitan France) and 1P (some Swiss cantons)

		5 years old	*6 years old*	*7 years old*	*8 years and over*
France	*1. N=930,100*	*2.1*	*82.2*	*13.4*	*2.3*
	2. N=871,600	*1.8*	*86.8*	*10.3*	*1.1*
Aargau	*1. N=5,768*	*0*	*0*	*85*	*15*
	2. N=6,230	*0*	*6*	*88*	*6*
Ticino	*1. N=3,419*	*0*	*92*	*8*	*0*
	2. N=2,665	*0*	*86*	*14*	*0*
Neuchâtel	*1. N=1,977*	*0*	*65*	*34*	*1*
	2. N=1,644	*0*	*62*	*36*	*2*
Geneva	*1. N=4,090*	*3*	*76*	*20*	*1*
	2. N=3,913	*4*	*74*	*21*	*1*

Key 1.= academic year 1980/81
2.= academic year 1989/90 (for France 1988/89)

Exemption from age requirements

When the chronological age is the only criterion taken into consideration, it amounts to denying the existence of different rates of development. Although such a position cannot easily be defended, some education systems are particularly rigid. However, accepting that psychological development should have an influence on the course followed by the pupil does not in itself indicate the best pedagogical measure to take. Very broadly, three solutions are possible for the gifted: going into a higher class, attending a school for gifted children or following enhanced courses (at school or outside of school). Since I cannot make use of comparative studies conducted in France or Switzerland, my aim here is not to assess these different proposals but to make a few observations on the question of being ahead in one's education.

In France, it seems that for a number of years, early entry into CP has been encouraged in favour of skipping a class during primary school (circular IV - 68-397 of 10 October 1968, quoted by Honoré, 1970). In the same article, Honoré shows, using national figures as well as his own studies, that young pupils do not repeat a year any more often than others, that subsequently they are directed towards the more prestigious fields and that they have better school results. However, a recent article in Le Monde de l'Education (Valo, 1989) stresses, on the basis of interviews with various experts, that exemption from the age requirement is not always risk-free and that the fear of having a child behind in his education leads some parents to regard it as a kind of insurance in case of his/her having to repeat a year.

French statistics show that being ahead in one's education was much more frequent 30 years ago (in 1960/61, 20.1% of children aged five attended CP, compared with 1.8% in 1988/89); however this information must be interpreted by comparing curricula, syllabuses and the way in which schools functioned at these different periods. It is highly probable that requirements have become more stringent with time. However, the system has not simply become more demanding since the statistics also show that the measures taken to deal with poor performance at school have been beneficial (rate for repeating a year in 1970/17: 17.5; in 1988/89: 10.1). According to A.M. Chartier, a French expert in the history of education, one would have to compare the criteria by which a child was considered good enough to skip the CP in 1960 with those of today. Her theory is that since in the 1960s the emphasis was placed on decoding skills, it was not desirable for teachers to be saddled with children who already knew how to read whereas now, with greater importance attached to comprehension, learning to write, and learning logic and mathematics, it is in their interests to hold onto good pupils who are thought to serve as examples for the others.

In Switzerland, where state education is concerned, the situation varies considerably from one canton to another. In a few cantons (including Aargau and Ticino mentioned earlier), a child is not allowed to advance a class. In many other cantons, a child may be permitted to be up to four months ahead (so-called "simple" exemption provided for in the Agreement approved by the confederation in 1970) of the earliest date of entry into compulsory schooling and skipping a year is sometimes authorised. However, these measures, which are regarded as very exceptional, are not always regulated (for further details see Bauer-Gauss, 1990). In Geneva, exemption from the age requirement of one year or more was authorised from 1P as from 1949 then regulated in 1961; they were recently (1988) extended to the 2ième Enfantine. Of all the Swiss cantons, skipping a school year is most widely practiced in Geneva. For this reason, but also because I have studied the situation in Geneva, I shall dwell on it somewhat.

Situation in Geneva

I shall confine my analysis to children skipping 1P, who were until 1988 the youngest children to benefit from an exemption from the age requirement. In this case, the regulations in force in Geneva provide that "exemptions from the age requirement of one year or more may be granted to children of compulsory school age who, from an educational, psychological and medical point of view are deemed fit to attend without difficulty a class normally aimed at older children" (regulation governing exemption from age requirement of 12 June 1974). Although I was responsible for the psychological examination for many years, my purpose here is not discuss the content of this examiniation (on this subject see Rieben, 1980) but rather to present some of the results gathered from the longitudinal study conducted between 1974 and 1977 and the study still in progress begun in 1979.

Table 2 shows that between 1974 and 1988 425 children were put forward as candidates (these figures only concern state schools and account on average for less than 1% of the population concerned), and the average rate of success was 61%. The number of children for which exemption is requested has risen slightly. The drop that occurred between 79/83 and 84/88 is only an apparent one and can be explained by the fact that as from 1986 roughly 50 children were admitted a year early into the 1P (this system came under regulation in 1988). Moreover, during the same period, school population dropped slightly.

Table 2

Geneva - Early entry into second grade
Number of candidates for the examination and success rate between 1974 and 1988

	74-78	*79-83*	*84-88*	*Total*
N children	*134*	*153*	*139*	*425*
success rate (%)	*59*	*68*	*55*	*61*

The figures in table 3 consistently show that more girls are entered for the examination for exemption from the age requirement than boys and that, between 1984 and 1988 they were more successful in this examination.

Table 3

Geneva - Early entry into second grade
Proportion of candidates and succes rate in the examination according to sex

	74-78	*79-83*	*84-88*	*Total*
% boys	*45*	*46*	*45*	*45*
% girls	*55*	*54*	*55*	*55*
% boys who passed	*58*	*57*	*44*	*53*
% girls who passed	*59*	*60*	*63*	*61*

Concerning the subsequent success at school of children exempted from the age requirement, the results of our first survey showed that 100% of the children relocated one, two, three or four years later (52 of the 59 children who took the exam), had remained ahead in their schooling. The second survey (see table 4), which analysed the situation at school five years after the examination, namely at the end of primary school, revealed similar results (only two children from the 1980 group repeated a year).

Table 4

Geneva - Progress between second and sixth grade (children having been exempted from the age requirement between 1979 and 1982)

	N at time of exemption	***N 5 years later***	***% in 6ème primaire***
1979	*24*	*17*	*100*
1980	*21*	*17*	*88*
1981	*22*	*18*	*100*
1982	*19*	*18*	*100*

With regard to the standard of academic results assessed on the basis of annual averages, our first survey showed that children exempted from the age requirement generally remained excellent pupils, were better than a control group which had not asked to be exempt from the age requirement and better than the group of children which had been refused exemption. According to the second survey (see table 5), which followed the pupils for a longer period to the end of their primary schooling, 88% of exempted children were either good or very good pupils assessed on the basis of their marks in French and mathematics. In comparison, only 50% of pupils who had been refused exemption were in the same categories.

Table 5

Geneva - achievement in sixth grade (as a percentage) (children applying for exemption between 1979 and 1982 and monitored 5 years later)

	Very good	***Good***	***Average***	***Poor***
Exemption granted N=70	*68%*	*20%*	*9%*	*3%*
Exemption refused N=25	*24%*	*26%*	*36%*	*24%*

On the basis of these two studies, it is possible to conclude that skipping the first primary class is a measure that can be an appropriate one, since our results reflect those obtained by Honoré (1970) for France. We can therefore conclude that a good psycho-pedagogical examination of the child that takes account of the circumstances surrounding the application for exemption contributes in theory to the right decision being taken. However, this solution remains expensive and, as far as Geneva is concerned, essentially limited to children from higher social backgrounds whose parents feel sufficiently comfortable with the education system to dare and to know how to ask for an exemption from the age requirement and to provide appropriate assistance outside school. If we approach the matter with the aim of giving all children who are ahead in their development the schooling that is best suited to their intellectual level, then this solution is not really sufficient. Perhaps the aim for the future would be to encourage differentiated education within the same class.

DIFFERENTIATED EDUCATION: A SOLUTION FOR THE FUTURE?

The problem of the schooling of gifted children can be examined in connection with the more general question of adapting the school to the rate of the child's development and to individual differences. This is the idea of the "made-to-measure school" put forward, among others, by Claparède as far back as 1920 and now discussed under the heading of "differentiated education" (see, e.g., Meirieu, 1985). It is important to discern two aspects of the question: why should education be differentiated, and how should it be done? As for the why, the fact that psychologists and educationalists have on many occasions observed individual differences in children of the same age and the fact that substantial savings could be made if primary schooling were reduced by 1-2 years for a considerable number of children, should suffice to justify differentiated education. However, the question remains essentially political and concerns the role assigned to the school: does it want to (and can it) at the same time try to attenuate individual differences arising from various sources (genetic make-up, family, social/cultural background, region, etc.) and at the same time allow each individual to develop to his or her optimal potential?

The question of how education is to be differentiated is more strictly psycho-educational. I have previously (Rieben & Foglia, 1986; Rieben, 1988) noted that in practice such differentiated education is often limited to declarations of intent in official texts. However, teachers are not to blame. There are prerequisites, both theoretical and practical, to differentiated education which are not fulfilled. The following steps would have to be taken: 1) a differential psychology of education would need to be developed, which for the moment is barely beyond the drawing-board stage, and which aims to identify differences in terms of aspirations, social and intellectual development, linguistic skills and learning strategies which are relevant in the analysis of learning processes at school (see for example Cronbach & Snow, 1977; Snow & Lohman, 1984; Entwistle, 1981; Resnick & Neches, 1984); 2) know-how and technical means enabling teachers to identify these relevant differences should be developed; 3) teachers should be encouraged to be more flexible in their teaching approaches.

In this more general context, the question of intellectually-advanced children could in most cases be resolved smoothly without the children having to skip a year but by allowing them to advance within a given course and within the class. This is a solution which appears to be taking shape in France (see document from the Ministère de l'Education Nationale. 1991). The "cycle" system which is recommended "should allow the pupil to take up, in the middle of an academic year, activities normally carried out in the following year, without changing class...Thus, early entry into the "elementary school" would no longer be necessary. Insofar as the upper section of nursery school belongs also to the basic learning cycle, it should begin to structure this learning as soon as the child is capable" (p.10). Research on this theme has also been conducted in Geneva by the Service de la Recherche Pédagogique (Hutin, Lurin & Soussi, 1990).

Bearing this in mind, psychologists should in the future focus their research less on finding more sophisticated psychological examinations in the strict sense of the term and more on seeking criteria that can be observed by teachers and that will allow them to regulate pupils' individually-adapted passage through school. In this way, each pupil, regardless of whether he/she is behind or ahead, could be placed in the most suitable pedagogical situation, while at the same time remaining a part of a class group, which can be regarded as a socialisation objective. With reference to the idea, presented above, of explaining individual variabilities in the context of general development models, it must be accepted that a joint study of pupils who are behind and those who are ahead at school will be more fruitful than two separate studies. This poses a veritable challenge for future psycho-educational research.

References

Aubert, A. (in preparation). *Développement de l'intelligence: Etude comparative d'enfants surdoués et d'enfants normaux, appariés par âge chronologique ou par âge mental.* Université de Genève: thèse en cours.

Bauer-Gauss, C. (1990). *La dispense d'âge d'une année au moins dans les écoles publiques genevoises.* Université de Genève: Mémoire de licence en sciences de l'éducation non publié.

Chauvin, R. (1975). *Les surdoués.* Paris: Stock

Claparède, E. (1920). *L'école sur mesure.* Lausanne: Payot.

Coriat, A. (1987). *Les enfants surdoués - Approche psychodynamique et théorique.* Paris: Païdos/Le Centurion.

Cronbach, L. & Snow, R. (1977). *Aptitude and instructional methods: Handbook for research on interactions.* New York: Irvington.

Entwistle, N. (1981). *Styles of Learning and Teaching.* Chichester: Wiley

Feldman, D.H. (1982). (Ed). *New directions for child development, 17, Developmental approaches to giftedness and creativity.* San Francisco: Jossey-Bass.

Freeman, J. (1985). (Ed). *The psychology of gifted children: Perspectives on development and education.* New York: John Wiley.

Gibello, B. (1976). Dysharmonie cognitive. *Revue de Neuro-Psychiatrie Infantile, 24,* 439-452.

Guilford, J.P. (1950). Creativity. *American Psychologist, 5.* 444-454.

Guilford, J.P. (1967). *The nature of human intelligence.* New York: McGraw-Hill

Honoré, S. (1970). Un an d'avance ou âge normal? *B.I.N.O.P., 26,* 245-253

Horowitz, F.D. & O'Brien, M. (1985). (Eds). *The gifted and talented - Developmental Perspectives.* Washington, DC: American Psychological Association.

Hutin, R., Lurin, J. & Soussi, A. (1990). *Approche qualitative de la division élémentaire: La première année enfantine - La première année primaire.* Genève: Service de la Recherche Pédagogique (document 90.3).

Inhelder, B. (1963, 2ème édition). *Le diagnostic du raisonnement chez les débiles mentaux.* Neuchâtel: Delachaux & Niestlé.

Lautrey, J., de Ribaupierre, A. & Rieben, L. (1981). Le développement opératoire peut-il prendre des formes différentes chez des enfants différent? *Journal de Psychologie Normale et Pathologique, 4,* 421-443.

Lautrey, J., de Ribaupierre, A. & Rieben, L. (1987). Operational Development and Individual Differences. In E. de Corte; J.G.L.C. Lodewijks; R. Parmentier & P. Span (Eds.). *Learning and Instruction. A publication of the European*

Association for Research on Learning and Instruction. (p.19-30) Oxford/Leuven: Pergamon Press/Leuven University Press.

Lieury, A. (in press). *Mémoire et réussite scolaire.* Paris: Bordas.

Meirieu, P. (1985). *L'école mode d'emploi - Des "méthodes actives" à la pédagogie différencié.* Paris: ESF.

Ministère de l'Education Nationale de la Jeunesse et des Sports. (1991). *Organisation de l'école primaire en cycles pédagogiques.*

Moser, U. (1989). *Hochbegabt-begabt-unterfordert: Erfahrungen aus dem Schulversuch "Überspringen einer Klasse".* Bern: Amt für Bildungsforschung der Erziehungsdirektion des Kantons Bern.

Resnick, L. & Neches, R. (1984). Factors effecting individual differences in learning ability. In R.J. Sternberg (Ed.). *Advances in the psychology of human intelligence* (Vol. 2, p. 275-323). Hillsdale, N.J.: Erlbaum.

Reuchlin, M. (1969). *La psychologie différentielle.* Paris: PUF.

Rey, A. (1963). *La connaissance de l'individu par les tests.* Bruxelles: Dessart.

Rieben, L. (1978). *Intelligence et pensée créative.* Neuchâtel et Paris: Delachaux et Niestlé.

Rieben, L. (1980). Psychologie de l'enfant et différenciation de l'enseignement dans la problématique des dispenses d'âge. *L'Orientation scolaire et professionnelle, 9,* 161-180.

Rieben, L. (1982). Processus secondaire et créativité: partie émergée de l'iceberg? In N. Nicolaidis & E. Schmid-Kitsikis (Eds.), *Créativité et/ou symptôme.* (p. 97-111). Paris: Clancier-Guenaud.

Rieben, L. (1988). Un point de vue constructiviste sur la pédagogie de maîtrise. In M. Huberman (Ed.). *Assurer la réussite des apprentissages scolaire? - Les propositions de la pédagogie de maîtrise.* (p.127-154). Neuchâtel et Paris: Delachaux et Niestlé.

Rieben, L. & Mengal, P. (1977). Intelligence globale, créativité et opérativité chez l'enfant: Analyse factorielle et analyse discrimininante. *Revue Suisse de Psychologie pure et appliquée, 36* (2), 100-108.

Rieben, L., Barbey, C. & Foglia, D. (1985). Différenciation de l'enseignement et apprentissage de la langue écrite. *Education et Recherche, 7,* 3, 29-50. Erratum, (1986), *8,* 2, 136.

Rieben, L., de Ribaupierre, A. & Lautrey, J. (1985). Le fonctionnement cognitif d'adolescent fréquentant des écoles de formation préprofessionnelle. *Revue Suisse de Psychologie pure et appliquée. 44,* 119-133.

Rieben, L., de Ribaupierre, A. & Lautrey, J. (1990). Structural invariants and individual modes of processing: On the necessity of a minimally structuralist approach for educational sciences. *Archives de Psychologie, 58,*29-53.

Snow, R.E. & Lohman, D.F. (1984). Toward a Theory of Cognitive Aptitude for Learning from Instruction. *Journal of Educational Psychology, 76,* 347-376.

Stamm-Riessen, M. (in preparation). *Hochbegabtenförderung in den Volksschulen der Deutschschweiz.* Université de Zürich, Thèse en cours.

Stednitz, U. (1989). *Begabtenförderung in Schule und Familie.* Schriftenreihe SFK/EPE/SPB, 10. Horw: Stiftung "Schule für das Kind".

Sternberg, R.J. (1985). Cognitive development in the Gifted and Talented. In F.D. Horowitz & M. O'Brien. (Eds.). *The gifted and talented - Developmental Perspectives.* Washington, DC: American Psychological Association.

Terman, L.M. (1925). *Genetic studies of genius*: Vol. 1. *Mental and physical traits of a thousand gifted children.* Stanford, C.A.: Stanford University Press.

Terrassier, J.M. (1981, 2ème éd. 1989). *Les enfants surdoués ou la précocité embarassante.* Paris: ESF.

Valo, M. (1989). Un an d'avance, est-ce utile? *Le Monde de l'Education, 159,* 22-25

Zazzo, R. (1956). La notion d'hétérochronie dans le diagnostic de la débilité mentale. *Revue de Neuro-Psychiatrie Infantile, 13,* 241-246.

Problems of the education of gifted children in the countries of Middle East Europe

Andrzej Sękowski
Poland

INTRODUCTION

Growing interest among psychologists, pedagogists and teachers in the problems of giftedness is due to the perception of the continuing shortcomings in the field of educating and upbringing gifted persons. In realizing the need to improve the methods of selecting gifted persons as well as the methods of working with them, specialists from many different fields are now seeking theoretical solutions which could be applied in educational practice.

Is there a specific approach to the problem of gifted persons in the countries of Middle East Europe, or is the situation there similar to what we observe in the United States and the countries of Western Europe? To answer this question one should pay attention not only to the research results in the field of giftedness, but also to the political and economic conditions which shape human attitudes and preferences. The political system which has been governing Middle East Europe since the Second World War has not positively influenced the atmosphere around the research on giftedness or on the educational work with gifted persons. The idea of egalitarianism was interpreted by Middle East European governments very superficially and instrumentally. So, instead of inspiring research and action in favour of gifted persons, it had the reverse effect. Political goals proposed by the persons in power were concerned with ensuring equal educational conditions for people originating mostly from the environment of workers and farmers. Particularly good conditions were created for those who identified themselves with the official ideology of the governments and people in power. Education of gifted persons was, unfortunately, often neglected and treated as a luxury, and sometimes viewed as harmful ideologically. As a result of such educational policy, many possibilities connected with the development of outstanding individuals were lost. Official opinion was that gifted persons could manage on their own, whereas to introduce such a category of concepts into science might have implications discordant with the ideology imposed.

The economic situation was not favourable for the development of research on giftedness or for the implimentation of programs in favour of gifted persons. Moreover, social feelings were not always in agreement with the ideas propagated by psychologists regarding the utility and even the duty to work for the development of giftedness and for the formation of the conditions to maximize this development. Research by social psychologists (Allport, 1954; Newcomb, 1965; Yuker, 1988) indicates that regardless of geographical location or political system, there is always a tendency in societies to develop negative attitudes towards minority groups, which also include disabled persons. Dissimilarity, i.e., deviating from social stereotypes, is the cause of the formation of negative attitudes. We also encounter such attitudes in relation to gifted persons. These are not manifested attitudes, but from time to time they are expressed in the decisions concerned with funding research or with the organization of the educational system. Furthermore, research carried out by social psychologists indicates that the mechanism of cognitive dissonance (Festinger, 1957) is induced by the perception of dissimilarity in the outer environment. This dissimilarity, discordant with the content and form of internalized cognitive structures induces the state of unpleasant stress. Such stress may be discharged by the application of a variety of mechanisms such as the attribution of negative features to gifted persons, looking for the causes of their success in external conditions, various types of privileges or facilitations, or in minimizing these successes. Another such mechanism is a tendency to negate the existence of giftedness as a subject of psychological research. Psychological analysis of the negative attitudes towards gifted persons basing on attribution theory (Heider, 1958) provides some understanding of the phenomenon. The tendency to explain the phenomena taking place around us is the result of a natural need to understand the causality of events. Where, then, is the cause of the fact that some people are more gifted than others, are more intelligent, have greater success in education, professional work, etc.?

Generally, the observer attributes these causes in the personality of gifted persons. At the same time the observer attributes excessive ambition to the gifted.

The problem of social attitudes towards gifted persons and policies in their favour is one of the key problems in the process of explaining the tendencies occurring in this field in the Mid-East Europe as well. It seems that the changes in human mentality, greater consciousness of the value constituted by human giftedness, are moving in the right direction.

In recent psychological and pedagogical literature, the issue of educating gifted girls (Borland, 1989; Sękowski, 1990) is often examined. It appears that sex may constitute a barrier to success in the life of gifted persons. It concerns both the West-European and Middle East European countries, as well as other countries of the world. It seems that in the Middle East European countries these barriers may be even more visible. An analysis of this situation suggests that one of the reasons for the formation of such barriers is created by the perception of gifted boys and girls by the social environment. The attribution of certain feminine and masculine roles in society favours the establishment of stereotypes. The results of research are sometimes interesting but they are not conclusive. For example, it appears that impulsive gifted girls are more positively perceived in the school environment than reflective gifted girls. Social perception of gifted boys is different. Impulsive boys are clearly underestimated, whereas reflective boys are perceived in a more positive way (Sękowski, 1990). This is an important point, since the position of a woman in the countries I am writing about is much more influenced by stereotypes connected with the attribution of certain roles to women, which reduce their possibilities for optimal development and realization of their own ambitions. On the other hand, there are no limitations of an organizational nature which restricts their equal chances for professional carrier. So it appears that the psychological barriers are often more difficult to overcome than economic, political or organizational ones. Particularly in fields such as mathematics, physics, and other sciences, there are more boys with recognized achievements than girls. In the field of humanities and fine arts this ratio is slightly different but even here there are more boys than girls.

The aim of the rest of the paper is to describe some trends in the research on giftedness in the countries of Middle East Europe as well as selected information concerning specific practical actions for the benefit of exceptionally gifted children. I do not aim at providing a complete review of all literature, but just to present some selected, characteristic forms of work and research trends. I will focus on the work of Polish scholars dealing with the subject of giftedness and discuss the aspects of this work which are also found in other countries of Middle East Europe. The political and economic changes that are rapidly occurring in this part of Europe makes it difficult to speak generally in terms of the old division of Europe into East and West, or Middle East. In the past this division was made for purely political reasons. Countries such as Czechoslovakia, Bulgaria, Romania, Poland, Yugoslavia have their specific cultural and historical individuality, and attempts to find common trends in the study of giftedness is a very difficult task.

RESEARCH TRENDS IN POLAND

Fundamental psychological research on giftedness in Poland is concerned with the structure of mind, intelligence and abilities, as well as with personality and environmental precursors of outstanding achievement. Such research is mostly carried out in university centers like Jagiellonian University in Cracow, Maria Curie-Sklodowska University in Lublin, Poznan University, Wroclaw Technical University. Of particular interest is the research concerned with creativeness and preferences in the field of cognitive styles. A number of valuable theoretical analyses of these topics have been made, for example, in Renzulli's Three-Ring Conception of Giftedness, Sternberg's Triarchic Theory of Intelligence, and Mönks' Triadic Interdependence Model of Giftedness. Given the variety of theoretical positions in the psychology of giftedness, it would be premature to speak of a dominating theory of particular value. The reservations raised by American and West European scholars about traditionally applied diagnostic techniques, e.g., intelligence tests which is more a measure of convergent than divergent thinking, often do not account for personality factors. Here, the contacts between Polish and American and Western scholars seem to be very important.

The research carried out in the Department of Psychology of Individual Differences, Warsaw University, under supervision of J. Strelau, deserves particular attention. His associate, I. Borzym had for many years directed the research concerned with the cognitive factors which underly giftedness and outstanding achievement. This research examined the capability for convergent and divergent thinking as well as the preferences in the field of cognitive styles. They mostly concentrated on the connection between the family situation of gifted children and their professional and academic achievements, and tried to account for both personality and environmental factors. The research was concerned with giftedness in mathematics, engineering, humanities and fine arts. By examining groups of people at different ages they could also examine the developmental dynamics of these factors. Some of the interesting results of this research are as follows:

- At younger ages there is relatively poor differentiation between convergent and divergent thinking, as well as in the preference of cognitive styles regarding the type of giftedness (mathematical, humanistic, artistic, musical). It does not only concern the preference for the cognitive style reflexiveness - impulsiveness.
- The level of general intelligence and the level of creative giftedness have an optimizing influence on the development of special talents. This relation is mutual, with changes in one type of giftedness resulting in changes in other types.
- The preferences in cognitive styles act as controls, determining the direction and form of activation of giftedness. It appeared that giftedness for sciences is more closely bound to reflexiveness and field independence, whereas giftedness in the humanities and arts are bound to impulsiveness and field dependence. However, it does not concern most creative persons.
- Cognitive style, as a formal feature of mental structures, may - within one category of giftedness - favour the formation of different methods of work.
- At the level of academic education, the gift for convergent thinking is the most valued and most highly assessed. It is clear that the gift for divergent thinking, i.e. the level of creative giftedness, receives too little attention.

Practical conclusions from this research, apart from the cognitive value, also have a functional value. They may be used by teachers and other persons taking part in educating gifted individuals.

In Poland research on giftedness is also being carried out by E. Nęcka, at Jagiellonian University, who is mainly concentrating on the issue of creative giftedness and creative thinking. At the Maria Curie-Sklodowska University, S. Popek's group is carrying out research concerned with the structure of giftedness, particularly artistic giftedness. A. Sękowski, is dealing with the effects of personality factors on achievements in various fields of human activity. He is also carring out research concerned with the social functioning of gifted persons, specifically with the musically gifted. W. Panek, Warsaw University (Bialystok Branch) is also dealing with the issues of giftedness. These individuals, as well as the many other scholars in this area, attest to the fact that interest in the problems of giftedness is clearly growing.

EDUCATIONAL DEVELOPMENTS IN POLAND

One of the most attractive organizational forms of the educational system for gifted children in Poland is the initiatives undertaken by the Polish Children's Fund in Warsaw. For several years the Fund has undertaken initiatives aiming at helping gifted children. The need to create optimal conditions for development of these children is becoming more and more evident. Gifted children are aided by the Polish Children's Fund with scholarships, which in addition to improving their economic conditions, also gives them the opportunity to make contacts with eminent scientists, scholars, artists and authors. While still attending elementary or secondary schools, they can participate in academic classes, sometimes starting their academic studies before they complete their secondary education, by following an individual program. Scholarship holders are classified into a few categories with regard to the nature of their giftedness. The basis for this criteria is constituted by the cognitive, humanistic, technological and artistic gifts. The holders of scholarships include prize winners of national competitions in various fields of knowledge, and participants and prize winners of international competitions. This classification, though deficient, allows to undertake a systematic attempt at improving the methods of work with students as well as to organize a system of knowledge acquisition for them and inspire them to undertake independent work. The activity of the Fund is particularly important for the students coming from small centers, towns and villages, since the possibilities of such students to have access to the most valuable sources of knowledge and research centers are quite limited. Such students, in spite of the technical and organizational problems which restrict the development of their gifts, manage in a very short time. The guiding idea of organizing help for gifted children is to include them in the general educational program. They are not separated from their social environment. From the point of view of the aims of comprehensive education, this approach is beneficial for the students. It seems that the poor conditions for development of the children coming from rural or small town communities is characteristic of the countries I am writing about. Poorer infrastructure, difficulties in transport and information transfer, and other differences in the level of education as a result of the poor rural environment, account for why the children in big cities have easier and more brilliant careers, and why they can develop their talents.

One of the forms of helping gifted children is constituted by camps organized by the Polish Children's Fund. This form of work is presently very popular in the United States and the countries of Western Europe. This type of work intensifies the educational process. Furthermore, in Poland, the participation in a camp for gifted children makes it possible to participate in classes held at the highest level. The lecturers are eminent representatives of different fields of knowledge. In a way, there exist two concepts of organizing such camps. One of them stresses education and development of particular special gifts. In this type of camp, e.g., for mathematicians, physicists, artists or musicians, the program is oriented towards a specific field and excedes substantially the educational program at school. The other type of camp attends to the general development of the students. In such camps, the participants are scholarship holders in various fields. The aims of such camps, beside the improvement of special gifts, are concerned with improving and enriching personality, broadening interests and improving activity. In addition to the educational value of the classes held, there are many social advantages. Making contact with eminent scholars goes together with the possibility of future co-operation with other scholarship holders of the Fund. It seems that the development of special gifts constitutes the principal idea of these camps, but the possibilities for the general psycho-social development of gifted persons is no less important.

Beside the activities of the Polish Children's Fund, other Foundations are also being established which aim at helping gifted children. This is partly the result of a general tendency where non-centralized enterprise and initiative in economic and political action is valued. Moreover, the issue of reforming the educational problem is more and more often raised by both parents and teachers. Education should be individualized, stimulating self-dependent intellectual work and original artistic activity.

Artistic schools have been particularly successful in educating gifted children and youth in Poland. It is the field where the effects of the extensive psychological research are most clearly seen. The system of educating musically gifted children allows for a relatively early and correct qualification of gifted persons for musical schools. Musical giftedness develops relatively early and improves mainly as a result of contact between a teacher and a student. Children are admitted to a musical school at seven years of age. The admission is preceded by a selection examination, which includes both a test for musical giftedness as well as a psychological test for school maturity. The examination is often preceeded by a few-days classes held by the teachers of musical schools. The children admitted to such school begin their music lessons and follow a general education program identical with that in standard schools. The process of musical education is divided into two stages. The first six years of learning constitute the primary musical school, and the next six years the secondary musical school. The students are educated practically, i.e. playing instruments, and theoretically, e.g., theory of music and history of music.

This system allows for a relatively quick development of the students' capabilities. The situation is similar with the process of general education. In musical schools the classes are relatively small, which means that the level of general education at the primary musical schools is high. Outstanding students can quite quickly participate in competitions and auditions of a test nature. During the earliest years the students learn to play piano or violin. During the later years of the elementary school, they begin to learn a second instrument. It is, however, a so-called optional instrument, i.e. the program of learning to play it is slightly limited.

The secondary level musical school aims at educating professional musicians. The general education program is slightly reduced; nevertheless, the graduates of this school are qualified enough to apply for admission to academic studies in other fields. The students of the secondary school participate in many auditions and competitions, to compare their skills with their contemporaries from other countries. It is interesting and characteristic of the system of musical education in Poland, that the general intellectual development as well as the development of personality of the students is strongly stressed there. The targets put forward by the educational system are becoming more pragmatic and the requirements for the students of musical schools will be concentrated more on the musical aspect than on general development.

In the field of psychology of music, valuable research is being carried out. It is concerned with psychological predictors of musical achievements (Manturzewska, 1974; Sękowski, 1989) as well as with the issues of the perception of music and learning. The Department of Psychology of Music, Warsaw Academy of Music, headed by M. Manturzewska (Manturzewska, 1974) adapted a set of tests examining musical gifts in various aspects to Polish environment.

The system of educating artistically gifted persons is also based on two levels of education. Secondary artistic schools provide students with a general education and artistic trade education. Higher levels of artistic education is provided in the Academies of Fine Arts and in Higher Artistic Schools. However, whereas the condition to be admitted to the Academy of Music is the completion of the secondary musical school, admission to the Academy of Fine Arts does not require completion of study in a secondary artistic school. There are artistic competitions organized for youth, and the prize winners often begin professional studies in this field.

There are many artistically gifted people who engage in artistic activities in a non-professional way. To help them is neither the target put forward by the system of artistic education nor by the Polish Children's Fund. However, there exists a system of the so-called Culture Houses, which participate in the process of artistic education of non-professional artists, as well as of those who are gifted.

The education of persons gifted for mathematics in Poland, Czechoslovakia, Hungary and the former U.S.S.R. is carried out in special education schools. A system of all-Poland competitions in mathematics provides the basis for identifying the most gifted students. There are special secondary schools with special programs in mathematics and physics. The educational program in these schools aims at an optimal development of the gift for mathematics. There are more hours of classes in mathematics and physics weekly than in other secondary schools. An expanded educational program is carried out and often the classes are conducted by research workers from universities. Exceptionally gifted students of mathematics and physics often take individual courses. An individual course of studies makes it possible to adjust the program of studies to the interests and gifts of the student. His contacts with the professors conducting the activities are held during their consulting hours. The tutors supervising such students try to compose the individual educational programs in such a way, that they agree with both capabilities and interests of the students.

The educational process for the persons gifted in the humanities is carried out at all levels of education. Among those persons, we can differentiate persons gifted for e.g. writing or language. Often these two skills are correlated. Language competitions are organized so that young people have an opportunity to demonstrate their skills and capabilities. In secondary schools there are programs specializing in the humanities. The program of subjects such as Polish, foreign languages, history is substantially expanded. Unfortunately, there are difficulties in the diagnosis of giftedness in the humanities. Sometimes, it is also difficult to establish appropriate criteria of evaluation of achievements in this field.

In general, the specific nature of various types of giftedness has an effect on the ease with which they can be detected. For example, giftedness in mathematics or music generally appears substantially earlier than literary or artistic giftedness. The difficulties in diagnosing special gifts often compel psychologists and teachers to apply the criteria of academic and school achievements. The tests for general intelligence, and the tests for creative thinking aid this diagnosis; nevertheless, as practice proves, they are not always diagnostic and prognostic with regard to gifted persons. The variety of didactic and auxiliary actions oriented towards gifted persons is an indication of the positive efforts being made. At the same time it can be seen that many possibilities for improving these forms of didactic and psychological actions still exist.

RESEARCH TRENDS IN OTHER EAST EUROPEAN COUNTRIES

The first specific research on giftedness in Czechoslovakia was carried out in the seventies. The research was concerned with sport (Bohemia) and intellectual capabilities (Slovakia). However, particular increase of the interest in this subject could be noticed in the eighties. In this period the Institute of Child Psychology and Psychopathology was founded - a group dealing with this subject. The research on intellectual, artistic and physical talents was undertaken, with the principal subjects of interest being the theoretical analysis of giftedness, identification of gifted persons, and the development of giftedness. Moreover, the research is concerned with creativeness, personality, and psychological methods for identifying gifted persons. There were also practical goals, such as the improvement of the education system for gifted persons. In other centers in Czechoslovakia research on this subject, with particular attention paid to the issue of creativeness, is also carried out.

Hungarian psychologists have substantial achievements in research on giftedness. Research carried out at the Hungarian Academy of Science concerning psychological conditions of giftedness and creativeness are well known. Psychologists, pedagogists, teachers and parents co-operate with one another. The achievements of Hungarian youth in mathematical and language competitions are also well-known. Hungarian psychologists are in close contact with the specialists from Western Europe and work very hard for quick integration of actions and research concerning giftedness carried out in Western and Middle East Europe.

Research on giftedness carried out in Yugoslavia has improved in recent years. In recent years research and publications of N. Jušovec (1986, 1988) or J. Makarowioć (1987) and others suggest that presently there exist positive tendencies for the development of research concerning gifted persons. Attempts at using L. Vygotsky's theory for research on giftedness are interesting. Presently, broad research is undertaken in many Yugoslavian research centers, such as at Zagreb, Belgrade, or Sarajevo. The subjects examined concern psychological diagnosis, pedagogical issues, educational process of particularly gifted pupils, psychological determinants of achievements in different fields of academic and professional activity.

The research concerning giftedness in Romania was mainly carried out by special pedagogists. The first publications concerned with the subject appeared as early as 1915. Although both the number and quality of those publications was not high, one can notice certain tendencies, which sometimes agree with the general tendencies in this field of knowledge. The results of research presented in Romanian publications are mainly concerned with the selection of gifted persons. They also deal with the issues of education of gifted children and their social functioning. Emphasizing the pedagogical aspect in research on giftedness also has a positive side to it. Maybe in the future it shall have a positive influence on the development of the psychology of giftedness in this country. Certain achievements in the field of aiding and developing talented children in this country should also be noted. Outstanding achievements in sports, e.g. in gymnastics, may serve as an example. Investigations of Romanian psychologists, who - in this way - could increase Romanian contribution in the education of gifted children and youth would be particularly important.

There is also a growing interest in the issue of giftedness in Bulgaria. Bulgarian specialists carry out interesting research concerning creativeness, cognitive styles, as well as psychological factors influencing outstanding achievements in various fields of activity.

THE ISSUE OF THE GIFTED DISABLED

One of the issues raised by psychologists, though perhaps not the most important one, is the education of gifted disabled persons. From the global point of view such research includes only a small group of people and perhaps therefore it is not worth while to pay too much attention to them. However, from the point of view of a psychologist, this issue is worth our particular attention. Disabled persons may constitute only a few percent of the population, but in absolute numbers this can be a substantial figure. The history of science gives us examples of gifted invalids, who - despite their handicap and certain resulting limitations - had eminent achievements in various fields of activity (e.g., H. Keller, L. Braille, M. Sizeranne). From the point of view of psychological research on giftedness, the results of investigations concerning disabled persons can be particularly valuable and may be treated as a topic for experimental research, e.g. examining the phenomenon of compensation of the sense of sight or touch also provides a better understanding of the mechanisms of intellectual functioning of a nonhandicapped person. Moreover, eminent achievements of disabled persons give evidence that it is not always the trait we accept as central, most important that decide about these achievements, but also other, less evident traits of personality. Such investigations are being carried out in Poland, and they should constitute a subject of interest for psychologists of giftedness and for persons dealing with aiding gifted persons. Gifted invalids have much more difficult conditions for fulfillment of their need for self-realization and optimum development. To create such conditions for them is a duty of civilized societies. These conditions cannot be the result of random decisions, but should result from scientific research and practical, considered decisions. Maybe in this field there is still much to do for the specialists from many different countries and parts of Europe.

Research on attitudes towards disabled persons supplies valuable information concerning psychological causes of the formation of prejudice toward every dissimilarity, such as towards minority groups and higher-than-average intellectual capabilities or achievements. Research on this subject is being continued and for many social and cognitive reasons should be continued all the time.

CONCLUDING REMARKS

Do the benefits to be gained in increased contacts between East and West go in one direction only? Can West European psychologists enrich their knowledge and skills as a result of contacts with psychologists from Eastern Europe? Doubtlessly, benefits resulting from such contacts can be mutual. However, why do only a relatively small number of papers written by specialists from Eastern Europe appear in the Journals published in Western Europe? The causes of this phenomenon are complex and it is difficult to provide clear answers. It is in any case clear that the period after the Second World War did not favour the making of permanent and useful contacts among scientists from different parts of Europe.

It would be very useful to organize common forms of work for gifted young people from different parts of Europe. Here I am thinking about common special camps as well as about international exchange in the field of academic studies. Moreover, it is very important for all researchers and practitioners dealing with the issues of giftedness to have an opportunity to exchange their experiences and research results.

A particularly important issue, which should be commonly undertaken by psychologists, is the diagnosis of giftedness. Although many different tests and psychological methods are in use, the level of such diagnosis is unsatisfactory. The predictive value of psychological diagnoses made on the basis of commonly used techniques is unfortunately very low. Those techniques which allow for intercultural research and may be applied to persons originating from different countries and environments are particularly valuable. Such diagnosis should include both cognitive elements (intelligence, creativity) and personality elements, as they are also very important, and sometimes they are more predictive than the traditional tests examining only the level of intellect.

The co-operation between specialists in the field of diagnosis and stimulation of giftedness may be mutually beneficial, and financial or economic factors should not play a decisive part here. Co-operation among future generations should be based on partnership relations. Our task is to create the conditions necessary for this process to be accomplished as soon as possible and in the most comprehensive way.

References

Allport, G. (1954). *The nature of prejudice.* Palo Alto-London: Addison-Wesley.

Borland, J. (1989). *Planning and Implementing Programs for the Gifted.* New York/London: Teachers College Press.

Festinger, L.A. (1957). *A theory of cognitive dissonance.* New York: Row Peterson.

Heider, F. (1958). *The psychology of interpersonal relations.* New York: Wiley.

Manturzewska, M. (1974). *Psychological determinants of success in music studies.* Warsaw: COPSA.

Mönks, F.J. & Ferguson, T.J. (1983). Gifted adolescents: An analysis of their psychosocial development. *Journal of Youth and Adolescence, 12*, 1-18.

Mönks, F.J. (1985). Hoogbegaafden: een situatieschets. In F.J. Mönks & P. Span (Eds.), *Hoogbegaafden in de samenleving.* Nijmegen: Dekker & van de Vegt.

Newcomb, Th. (1965). *Social psychology* (in Polish). Warsaw.

Renzulli, J.S. (1986). The three-ring conception of giftedness: A developmental model for creative productivity. In R. Sternberg & J. Davidson (Eds.), *Conceptions of giftedness.* Cambridge: Cambridge University Press.

Sękowski, A. (1990). The perception of gifted boys and girls at school as related to their reflective-impulsive cognitive style. Report presented at the Second European Conference on High Ability. Budapest.

Sękowski, A. (1989). Personality and artistic achievements of the students of musical schools (in Polish). Wroclaw, Warsaw, Cracow, Gdansk, Lodz: Ossolineum.

Sternberg, R., & Davidson, J. (1986). *Conceptions of Giftedness.* Cambridge: Cambridge University Press.

Sternberg, R. (1990). *Metaphors of Mind.* Cambridge: Cambridge University Press.

Yuker, H.E. (Ed.) (1988). *Attitudes toward Persons with Disabilities.* New York: Springer.

Concepts of giftedness, and research into the education of gifted children in the Netherlands

Pieter Span
The Netherlands

INTRODUCTION

Dutch research on gifted ability began with the appointment of G. Révész to the Chair of Psychology at the University of Amsterdam in 1932. Even by that time, Révész was a psychologist of note in Europe, most particularly as a pioneer of research on giftedness. In his original work he analysed biographies of famous gifted people, and reached the conclusion that "giftedness" is not a "general" attribute (Révész, 1946). Instead, he concluded that it is an innate potential for certain areas of learning such as mathematics, sciences, knowledge of men, performing arts, etc. In order to reach exceptionally high achievements, a gifted individual would focus specifically within each area, but the final outcome would be subject to the influence of an interaction of internal and external factors. Révész's outlook affected that of his students, notably A.D. de Groot, and in turn the next generation of psychology students, including J. Elshout and the present writer.

Révész and his colleagues introduced their new ideas about giftedness and the consequent selection of the gifted at a meeting of Dutch psychologists in

Amsterdam in 1938. H.C.J. Duijker, who was one of this team, specified the procedures of theory and practice - the analysis and selection of the gifted. In the first, the researcher aims for insight into the characteristic psychological structure of gifted individuals; in the second, the goal is to trace phenomena indicating dispositions which could be developed into gifted behaviour. Since progress made in the theoretical field would influence the applied activities, any fundamental problems would be more likely at the initial stage.

This paper discusses the two aspects of theory and practice in the Netherlands: first, the nature of giftedness, and second, the practical effects of these considerations and definitions on the identification of gifted pupils in Dutch schools.

THE NATURE OF GIFTEDNESS

Duijker (1938) believed that giftedness could only be judged by the quality of the ensuing product, and that such judgement must be made by specialists in the particular field. The suggested sequence was thus to look at the gifted achievement first, and only then at the psychological characteristics associated with it.

De Groot's (1938) research on gifted chess players considered these psychological characteristics. (His 1946 doctoral thesis on "Thought and Choice in Chess" became world famous). He discovered that of the characteristics which are necessary for gifted chess performance, considerable experience was top of the list. Of itself, experience enables a chess player to store moves and strategies in memory, but even more important is the fact that abstraction and subsequent generalisation can *only* come from the experience of practice tournaments. Playing chess at a gifted level demands both analogical reasoning and "thinking fantasy", that essential ability to perform "mental jumps" (in current terms: convergent and divergent thinking). De Groot argued the importance of willpower as a function of the urge to win - the major point of difference between giftedness in chess and mathematics.

After World War II, De Groot (1971, 1980, 1985) was the only psychologist in the Netherlands to continue research on giftedness, and he continued to publish work regularly about aspects of intelligence. His ideas remained consistent: that giftedness is a natural potential for something specific - it is in the genes - and that the level of ability should be judged by experts, based on the individual's achievement. However, he saw "general giftedness" as potential that could lead to high achievement in areas which were yet to be defined. General giftedness, though, depended on high intelligence, by which he meant the mental perception of problems and methods of processing not bound by context. Creativity and activity (motivation) are inherent aspects of intelligence; creativity being equivalent to divergent thinking, while activity is the natural need to use and develop one's "brains".

De Groot believed that because it is potential which imposes the ceiling on achievement, in principle, when conditions of upbringing are optimal, each individual can reach that personal ceiling. Thus, the more a society equalises life opportunities, the more people's manifest differences would be due to potential. He assumed that this was so for most of the Dutch population, so that individual differences in achievement must be due to differences in potential. However, there are distinct inequalities of life opportunities and great variations of opportunity in developing countries, and so there must be great differences in achievement from the same basic potential.

Using this principle, De Groot raised an emotional storm in the early seventies by calculating that in modern Western countries, about 80% of the total variance in intellectual achievement was due to inheritance. P. Vroon provided a typical response by considering the statistical problems in computing this proportion. He also denied De Groot's statement that for Dutch working-class children there were no longer any barriers for "hidden talent". Nevertheless, one finding of large-scale research on the intelligence levels of the Dutch population was that there was in fact very little "hidden talent" at low socio-economic levels (Van Heek, 1968). An alternative suggestion was that it was the children's social and physical environments which were not conducive to their taking part in school-type activities, a view in which Meijnen and Guldemond (1985) concurred on the basis of their own research findings. They studied about 500 gifted pupils and discovered a strong correlation between socio-economic status and level of secondary education; out of the high, medium, and low SES-groups, 83%, 76% and 61% respectively participated in the highest school level.

For other reasons, followers of the cultural-historical school originated by L. Vygotsky disagreed with De Groot (Blom & Wardekker, 1983). They were convinced that it was impossible to predict individual development to such a fine degree. Although they did not deny that differences in intelligence were present at birth, they saw its further development as very dependent on two aspects: the available meaning systems and, more importantly, on the way the individual is able to take in and adapt to those systems. This appropriation is, in essence, an active process of attribution of meaning. A relatively low level of either meaning systems and/or appropriation at birth can be compensated for in later life.

The consequences of this discussion have been mainly of an educational-political character (eg. delayed selection in the school system), and have had only modest influence on the analysis or measurement of giftedness. Dutch intelligence tests are still based on the classical test theory, and the test development for learning potential, linking up with Vygotskian theory, is only just starting (Span, 1988).

The cognitive psychologist, J. Elshout, has refused to be drawn into the nature-nurture debate, linking up instead with De Groot's original chess research by measuring achievements. He has identified quantitative and qualitative differences between the achievements of experts and others. Experts in any particular domain were found to have available both more relevant perceptual patterns and thinking schemata than ordinary individuals, and these patterns and schemata were of a higher level. These mental procedures are more geared to application and are also based on a deeper understanding of the subject matter. Boshuizen (1989) calls the schemata used by experienced medical doctors "illness-scripts". The expert is also able, depending on the difficulty of the task, to switch between implicit, automatic action and that based on explicit know-why (Elshout, 1991).

To attain a state of expertness, Elshout recommends a minimum of several thousand hours of training. The gifted, he says, are those who have the required personal qualities to do this: high intelligence, the liking for intellectual challenge, persistence, independence, and professional self-confidence (Elshout, 1983). Over the same lengthy time-span, they progress at a faster rate than others who are also strongly motivated. This difference refers to the aspect of ability, but because this is a construct, it is not directly measurable.

Is giftedness observable before the age of 15? Here too, Elshout (1988) refers to qualitative differences in perception and thinking. He terms children "expert-novices" who, when confronted with new and difficult tasks, do not show trial-and-error behaviour, jump to conclusions, or operate in an ill directed way, but instead work systematically and make good use of available information. Process analysis of the manner in which children solve problems could provide insight into the quality of those processes.

In fact, this was attempted by Span and Overtoom (1986, 1990; Overtoom, 1991), who studied two groups of children selected by very high and average intelligence test scores, backed by teacher nomination. The children were confronted with new and difficult mathematical problems, and their activities and manner of problem solving were then systematically analyzed within the framework of the "orientation - execution - evaluation" model. It was clear that there were considerable differences between the two groups with regard to the way they orientated themselves and evaluated their solutions. The more intelligent took sufficient time to analyze the problem, paid attention to several steps in the solving process, and showed a tendency to reflect on their own action, an approach consistent with Elshout's description of expert-novice behaviour.

Elshout (1988) has made a plea for reinforcing the construct of "intelligence" with metacognitive considerations. This would mean including individual access to cognitive order and structure, the use of general techniques (means-end analysis), construction of a usable memory-trace of the solving process, etc. Like Siegler and Kotovsky, Span (1988) emphasised the value of research on a new hierarchical system of levels of analysis by investigating the relationships, in both high and averagely intelligent individuals, between elementary information processing, combining and comparing processes, and metacognitive skills. At the moment, though, there is no sign of such research being undertaken in the Netherlands, nor even any possibility of (or finances for) fundamental research on intelligence.

There are, of course, non-cognitive components of gifted (expert) behaviour. Elshout (referring to Taylor & Barron) sees this mainly as a matter of skill and will. Yzerman (1970), assistant to Van Heek in his research on hidden talent in the Netherlands, also believes that personality characteristics like motivation, concentration, and persistence are essential for the attainment of very high achievements. Mönks (1991), basing his outlook on the work of Renzulli and Yzerman, sees the most important characteristic for excellent achievements as being motivation, given that the requisite high intelligence and creativity are present.

To summarise the findings thus far: There is strong agreement that giftedness can only be identified after the production of outstanding achievement, and that high intelligence and hard work are seen as conditional for such an achievement. The combination leads to "expertness". This makes it difficult to speak of "gifted children" because of their lack of time to become an expert. Therefore, it would be better to speak of highly able children, indicating that conditions are favourable for the development of gifted behaviour. High intelligence is the *sine qua non*, actualised in a rapid cognitive development, an advanced knowledge-base, and adequate metacognitive skills. These characteristics are open to psychological testing. But if the socio-cultural circumstances are unfavourable (eg. poor family, developing country), learning potential tests could provide the necessary indication.

THE IDENTIFICATION AND EDUCATION OF GIFTED PUPILS

Primary education for children from 4 to 12 years-old in the Netherlands takes place in mixed ability classes. But in secondary education there is both selection into different schools, as well as "streaming" into different classes (levels) within the comprehensive school. At the moment, secondary education starts with a "bridging" year, to give pupils the chance to prepare themselves for entering the highest possible level of education. A law is in preparation to extend this to three years, although gifted pupils would also be allowed to take the normal three-year curriculum in two years.

However, a high percentage of the best achieving children attend the selective gymnasium schools, although gifted children of low SES are still under-represented both there and in the upper streams of the comprehensive schools. In spite of delayed pupil selection and pressure by the government to amalgamate small schools, the gymnasium continues to be popular with parents. Although the Netherlands has no special schools for gifted pupils *de jure*, the gymnasium functions *de facto* as such.

After World War II, it was essential to direct almost all the nation's attention and resources into reconstructing the country, which limited the concern and finance available for education. It was only in the 1960s that serious attention was given to the special problems of gifted pupils in the schools (the panic in the USA about the launching of the Sputnik in 1957 adding an extra stimulus). In 1962, a Dutch-Belgian conference on the selection and education of gifted pupils concluded that special schools and special classes should be discouraged, but internal differentiation, by means of a system of tasks and a less stringent classroom organisation, should be encouraged. The British comprehensive school was accepted as the best model for secondary education.

At a 1963 symposium on creativity at the University of Nijmegen, Mönks proposed longitudinal research on the development of gifted children. And in the same year, a South African educational psychologist, studying in the Netherlands, published *Giftedness and the Education of Gifted Children* (Duminy, 1963). He described three educational options:

1. Types of acceleration (jumping classes or studying at a personal pace).
2. Enrichment, interpreted as well-directed thinking and study methods, as well as a capacity for judging, critical attitudes to learning materials, etc.
3. Homogeneous grouping, based not solely on IQ, but also on children's take-up of teaching, though many such grouping experiments failed because there was no appropriate teaching available.

By the mid-1960s it was possible in theory for every teacher in the Netherlands to be acquainted with the difficulties which could be experienced by gifted pupils. Information came, for example, from the Belgian psychologist R. Vandenberghe, who published two long articles in a popular Dutch educational journal between 1964-1966 about gifted pupils in American schools. Consequently, the Dutch Union of Teachers promoted the idea (1965) that gifted pupils should remain in normal schools but that teaching should be adapted to their special needs, with provision for possible acceleration and enrichment.

It was also in this fertile decade that Van Heek and Yzerman had done their research (above) on "hidden talent" in lower SES groups. However, highlighted by the 1968 Paris student revolt, feelings that everyone was now "equal" affected Dutch education, and the message from Yzerman's 1970 publication was neglected. So, around 1970, leading educationalists in the Netherlands agreed on the necessity of education adapted to highly able pupils. The favourite solution was enrichment, with an accent on the learning of critical thinking. Preferably, this should be accomplished in mixed-ability classes; acceleration was a good second choice. A combination of enrichment and acceleration was found in schools with more flexible classroom organisation.

Curiously enough, this movement in the eighties did not link up directly with Dutch "tradition", but was inspired by the situation in the USA.

The Dutch "counter-revolution" began in 1979 with the creation of the Dr. Binet Foundation. Its goals are: "The promotion of the guidance of the gifted, who because of their capacities, experience difficulties in society". It continues to offer advice to gifted children and their parents, and also organises courses for both parents and teachers.

Several things happened in the first half of the 1980s. The influential journal *Pedagogische Studiën* published a special issue on giftedness (Span, 1983), containing an analysis of the Dutch situation at that time, which put the cat among the pigeons (Mönks, 1983). A congress at the university of Nijmegen in 1984, on The Gifted in Society, brought together all those involved with research on intelligence and creativity in the Netherlands (Mönks & Span, 1985). The then State Secretary for Education took a personal interest in the adaptation of education to gifted pupils. She commissioned Mönks (University of Nijmegen) and Span (University of Utrecht) to investigate and then advise on the identification and education of gifted pupils in mixed-ability classes during the bridging period. This was the first empirical research in the Netherlands concerning gifted pupils in school.

The Utrecht research

This research was based on the "Amsterdam tradition" of giftedness, which recognises convergent and divergent thinking, and the acquisition of knowledge. About 940 pupils in the bridging classes of four schools were given a small battery of tests, not including personality. Teachers and pupils were interviewed, and several classes observed (Jansen Schoonhoven et al., 1986; 1987). Three groups of "gifted" pupils (42, 45, and 16 respectively) were distinguished by test results, by the teachers, and by both. There was clearly some disagreement. Teachers had distinct views about the first two groups, describing the test-nominated children as sloppy and lacking in concentration, but their own nominees as conscientious and attentive, fitting their image of the gifted pupil. However, three years later, hardly any children from these two groups had achieved really high marks in school. The test-nominated group were especially disappointing, but several un-nominated pupils performed extremely well. There should evidently be very great care with the use of these kinds of identification methods.

At the same time, a nationwide survey had found that although teachers were willing to teach the gifted in mixed-ability classes, they felt lacking in expertise. These combined results led Span and colleagues (Stevens et al., 1987) to suggest a change of policy for gifted secondary pupils to the State Secretary for Education. In summary:

- Good education for all pupils, including the gifted, is fundamental, but intelligence tests should only be used to identify suspected under-achievers.

- New measures should link up as closely as possible with existing school provision.
- Depending on the school's capacity to vary its teaching, enriching learning materials should be added appropriately, or better still, integrated into the regular curriculum.
- Enrichment materials should be developed, and teachers taught how to teach mixed-ability classes.

The Utrecht team was encouraged to continue along those lines because the State Secretary was successful in her parliamentary defence of the policy for the education of gifted pupils. The next part of the research began by posing three questions to a relatively large comprehensive school (Van Dijk, et al., 1990):

1. How is the school coping with pupils' different abilities?
2. What types of teaching are already in use?
3. What is the spread of intelligence among the pupils?

It appeared that, contrary to its stated aims, the school did not have genuine mixed-ability classes in the bridging period, but had streamed its pupils by ability on recommendations from the elementary school. This tendency, found in many comprehensive schools, arises because teachers cannot cope with different kinds of teaching in the classroom. (In Rotterdam, one school has a special bridging class for gifted pupils). The school preferred children to be streamed early, rather than moving them from one stream to another, and so the (few) gifted pupils had been congregated in one class.

Then, the team analyzed the school's small amount of differentiated teaching, i.e. special learning materials, learning time, and levels of processing. This choice of differentiated measures grew naturally from the familiar Dutch model for differentiated teaching, as below, where successful pupils in Test I went on to new content, and at a higher level of processing.

```
                      fail --- Repetition
                       /                  \
Basics -----> Test I                       ------> Test II
                       \                  /
                      pass --- Enrichment
```

To broaden the sample, two further schools were brought into the research, but very little of their enrichment material (native language and mathematics) met international criteria, so that almost all the materials for highly achieving pupils had to be newly designed by the team. And even worse, as neither was using the model in a systematic way, highly achieving pupils were not sure of being taught in an optimal manner. Hence, the team's advice was to adapt provision to each teacher's way of teaching. This resulted in improvement for gifted pupils in both the speed of teaching and the provision of supplementary enrichment. Most profitable, however, were tasks that could be processed by the pupils at several levels. Such "integrated enrichment" was used by pupils working in small groups at their own levels (Van Dijk, et al., 1991).

Consequently, considerable time was devoted to finding adequate materials for gifted secondary pupils. Meanwhile, the nationwide survey carried out in parallel, yet within the research, also found that teachers had very little material to work with in this area. The team then reviewed the most frequently used methods, and catalogued those which were the most appropriate for the gifted. This register, given in practical terms, together with directions on how to judge other teaching methods, is now available (Kok et al., 1991).

The Nijmegen research

Based on Renzulli's definition of giftedness, about 770 pupils in bridging classes were given a battery of tests and questionnaires, to assess intelligence, creativity, personality, achievement motivation and self-concept, as well as teachers' opinions and school-marks. Eighty gifted pupils emerged - more than 10% of the sample - and were divided into two groups on the criterion of accord between school achievement and intelligence scores; 75% were seen to be achieving satisfactorily, but 25% were underachievers. The latter group had significantly more negative attitudes about self-concept, school-work, and every kind of teaching (Van Boxtel & Roelofs, 1987; Van Boxtel & Mönks, 1991). A second analysis of 120 interviews found no differences in home and leisure activities between the two groups, but was able to reaffirm their attitudinal differences (Dijkman, 1991). The Nijmegen team endorsed Renzulli's methods for selection and education of the gifted: specifically that the most highly achieving pupils should be considered for acceleration and enrichment, with identification by a "talent search". Underachievers should be guided individually.

The members of the Nijmegen group changed their focus from secondary to primary education in 1986. They began by sending a questionnaire to 170 teachers about their attitudes towards and experiences with gifted children (Van Boxtel et al., 1987). The findings were that the teachers believed the term "gifted children" to mean the same as "child prodigies" in the area of intellectual achievement, that such children are vulnerable to social and emotional problems, and above all, that underachievers should receive additional care. Teachers preferred mixed-ability classes, and would avoid acceleration unless there is no alternative solution. On the basis of these results, the team suggested a need for identification methods, enrichment materials, and the avoidance of social isolation for the children. Frequent consultations with the school advisory service should also be encouraged.

Over the next two years, the search was directed to finding developmental structures to be used for the gifted in elementary schools. It used the Schoolwide Enrichment Model (Renzulli & Reis, 1985) in a pilot study in three schools. They aimed to find:

- Integrated measures for identification and education.
- Supporting activities for external guidance and counselling.
- An educational model and curriculum for dedicated teacher-training.

The work placed strong emphasis on identification, and resulted in the following suggestions:

- The main criteria should be intelligence and achievement. Other skills, such as creativity and motivation, could be assessed during enrichment activities.
- Identification should be carried out in two phases:
  (a) Pre-selection of the potentially gifted by an intelligence test and brief teacher checklists.
  (b) Assessment of the preselected children by a teacher questionnaire, other available information and, if needed, further diagnostic activities.
- Ultimately, only educational practice can show whether children are rightly or wrongly identified and, since no procedure or method is flawless, the procedure should be flexible and open to continuous reassessment and use of new methods.

The main recommendations from the research are concerned with the Schoolwide Enrichment Model. This is so difficult to realise, implementation taking from 3 to 5 years, that very few schools could manage this without help. The most important essential is that the teachers are willing to participate, and invest extra time and energy in it. The main conclusion was that the Renzulli system has value, but that it should not be considered as a blueprint for Dutch schools, instead can be used as a general guideline for building an enrichment model within a particular educational situation. The full report of this work is available (Haest & Sanders, (1991) in Platenkamp, 1991).

During this time, the Institute for Applied Social Studies, also at the University of Nijmegen, investigated ways of improving individualised learning for gifted children in schools, via relevant curriculae and additional learning materials (Mooy, 1991, appendix). Their conclusions were entered on a national computerised data-base, to which all 64 Dutch school advisory services have access. As of January, 1991, the IAS have started a cooperative project with the Institute for Curriculum Development in Enschede, to develop specialised individualised curriculae, eg. in native language and mathematics, due in September 1993.

Parallel to the research in Nijmegen, a separate study by the schools advisory service in the city of Gouda (Kort, 1989) involved four elementary schools. Ten per cent of the pupils were perceived as gifted, though this high percentage was strongly criticised: according to their teachers, 80% of these pupils feel happy at school, but the other 20% were in need of extra guidance. The researchers felt that giftedness should be identified as soon as possible, with as friendly a procedure as possible. The Gouda advisory service also developed enrichment materials, linked with the regular elementary school curriculum.

Other elementary school enrichment materials were developed by the advisory service of Zoetermeer, but these are not linked with the regular curriculum; they are extras such as the study of foreign languages, planets, Trojan war etc. The scheme was tried out in 7 schools, with 30 pupils having a WISC-RN IQ of at least 125 (De Bruin-de Boer et al., 1990).

All the findings reviewed here are be described and supplemented with recommendations in a publication by the National Pedagogical Centre (see Platenkamp, 1991).

In summary: both the Netherlands Ministry of Education, and school-teachers are in entire agreement that special education for gifted pupils should be given in mixed-ability classes. But at the same time it is apparent that teachers are inadequately trained to handle the necessary classroom organisation to carry it out. Moreover, the learning materials to which they have access cannot be used across a wide range of pupil abilities, notably the highly able. Hence, the two major priorities in the Netherlands should be for the development of appropriate learning materials, as well as for appropriate training for teachers. Without that impetus, the most obvious alternative is "streaming". Enrichment models, presently unknown in Dutch schools, could only be used as guides because any measures taken should be as close to daily practice as possible, because otherwise they are unlikely to be accepted or used.

OTHER PROVISIONS FOR GIFTED YOUTH

- Older pupils can take part in International Olympiads in mathematics, physics, and chemistry. They are first selected by the school, and the top few receive a short extra training course before they enter.
- No special arrangements for gifted students are available at university level, but at the end of their four years study the best achievers are considered for the post of "assistant in training". If accepted, they are involved in research and can present a dissertation, again after four years. However, only a small percentage of people can stay on at university afterwards. Very promising researchers can apply for governmental grants

(Huygens grants), and the Dutch Organisation for Scientific Research sometimes funds young researchers with "frontier modifying" grants.

- There are 230 music schools and creativity centres in the Netherlands, where even very young children can take weekly lessons in dance and music. There are also 11 conservatories, at which children can combine normal secondary education and music tuition; some take in gifted young musicians. But highly gifted children, and child prodigies, are more likely to be tutored privately, although this does depend on the finances and motivation of the parents. They can all enter the regular national competitions.
- Thousands of sports associations try to find and foster talent, although there is no specific identification procedure. Sports teachers are trained through the Academies of Physical Education and Sports Schools, but not youth. But there is one part-boarding school for gifted young athletes in the city of Arnhem (Papendal). In general, the route to the top, such as the Olympic teams, is expensive, packed with frustration, and many complain about what is (or is not) provided. But thanks to TV, soccer players such as Cruyff and Van Basten are known all over the world.
- The Association of Parents of Gifted Children, PHAROS, was founded in 1987. It has branches in about 20 local areas, and is very active, both educationally and socially.

References

Blom, S. & Wardekker, W.L. (1983). Hoogbegaafdheid: geen vanzelfsprekend begrip. *Pedagogische Studiën, 60,* 5, p. 220-229.

Boshuizen, H.P.A., (1989), *De ontwikkeling van medische expertise.* Maastricht: Dissertation.

Boxtel, H.W. van (1987), *Hoogbegaafden in het Basisonderwijs,* Deel II. Nijmegen: Center for the Study of Giftedness.

Boxtel, H.W. van & Mönks, F.J. (1992). General, Social and Academic Self-Concepts of Gifted Adolescents. *Journal of Youth and Adolescence, 21,* 2, 169-186.

Boxtel, H.W. van & Roelofs, J.J.W. (1987), De hordenloop van de begaafde onderpresteerder. *Didaktief, 17,* 5, p. 10-13.

Boxtel, H.W. van, Broekman, G.H.A. & Roelofs, J.J.W. (1987), *Hoogbegaafden in het Basisonderwijs.* Deel I. Nijmegen: Center for the Study of Giftedness.

Bruin-de Boer, A.L. de, Greef, E.E.M. de & Heuij, A.A. van den (1990). *Verslag over de tweede helft van het projectjaar.* Zoetermeer: Stichting regionale onderwijsbegeleidingsdienst Zuid-Holland West.

Duminy, P.A. (1963). *Begaafdheid en die onderwijs van begaafde kinders*. Amsterdam: Meulenhoff, Serie verdiept inzicht, nr. 18.

Duijker, H.C.J. (1938). Methodologische grondslagen van het onderzoek der begaafdheid. *Algemeen Nederlands Tijdschrift voor Wijsbegeerte en Psychologie, 32,* 2, p. 80-88.

Dijk, W. van, Kok, W.A.M. & Span, P. (1991). *Begaafde leerlingen in heterogene klassen*. Utrecht: ISOR.

Dijk, W. van, Kok, W.A.M., Olde Weghuis, J.M.M. & Span, P. (1990). *Schoolorganisatie en de opvang van begaafde leerlingen in het voortgezet onderwijs; de analyse van een beginsituatie*. Utrecht: ISOR.

Dijkman, T. & Venmans, A. (1991). *Begaafde leerlingen in het voortgezet onderwijs: Aspecten van hun leefwereld.* Nijmegen: Instituut voor Toegepaste Sociale Wetenschappen.

Elshout, J.J. (1988). Intelligentie en goed beginnen. In: G. Kanselaar, J.L. van der Linden en A. Pennings, *Begaafdheid: Onderkenning en beïnvloeding*. Amersfoort /Leuven: Acco.

Elshout, J.J. (in press). Talent: the ability to become an expert. In: J. Freeman, P. Span and H. Wagner (Eds.), *Actualising Talent: A lifelong Challenge*. Göttingen: Hogrefe.

Elshout, J.J., Kouwenhoven, K. & Tromp, D. (1983). De identificatie van hoogbegaafde jonge mensen. *Pedagogische Studiën. 80,* 5, p. 209-219.

Groot, A.D. de (1938). Begaafdheid voor het schaakspel. *Algemeen Nederlands Tijdschrift voor Wijsbegeerte en Psychologie, 32,* 2, p. 97-105.

Groot, A.D. de (1946; 1965). *Het denken van de schaker, Thought and Choice in Chess*. Den Haag: Mouton.

Groot, A.D. de (1971). Het eigen vooroordeel en de strijd daartegen, *Intermediair, 7,* 8, p.11 ff.

Groot, A.D. de (1980). Over de betekenins van intelligentieverschillen. In: A.D. de Groot en J.C. Traas (Eds.), *Onderwijs van binnen en van buiten*, Deventer: Van Loghum Slaterus.

Groot, A.D. de (1985). Over algemene begaafdheid: begrip, manifestaties, verdeling. In: F.J. Mönks en P.Span (Eds.), *Hoogbegaafden in de samenleving*. Nijmegen: Dekker en van de Vegt.

Guldemond, H. & Meijnen, G.W. (1985). Hoogebegaafden en het voortgezet onderwijs. In F.J. Mönks en P. Span (Eds.), *Hoogbegaafden in de samenleving*. Nijmegen: Dekker en van de Vegt.

Haest, M.C. & Sanders, M.P.M. (1991). *Implementation of the schoolwide enrichment model in elementary education in the Netherlands (Summary)*. Nijmegen: Center for the Study of Giftedness.

Heek, F. van (1968). *Het verborgen talent.* Meppel: Boom.

Jansen Schoonhoven, A., Linden, J. van der, Span, P. & Susante, M. van (1986). *Hoogbegaafde leerlingen.* Den Haag: Selecta Reeks, SVO.

Jansen Schoonhoven, A., Linden, J. van der, Posthuma de Boer, M., Span, P. & Susante, M. van (1987). *Hoogbegaafde leerlingen,* Deel II. Den Haag: Selecta Reeks, SVO.

Kok, W.A.M., Poorthuis, G.M.T. & Pool, I. (1991). *Additionele leermiddelen voor begaafde leerlingen in het voortgezet onderwijs: Criteria, beoordelingsinstrumenten, beschrijvingen.* Utrecht: ISOR.

Kort, H. (1989). *Projekt begaafden in de basisschool.* Gouda: Schoolbegeleidingsdienst Midden-Holland en Rijnstreek.

Mönks, F.J. (1983). Hoogbegaafde kinderen: een benadeelde groep? *Pedagogische Studiën, 60,* 5, p. 195-208.

Mönks, F.J. (1989). Hochbegabte in den Niederlanden: Forschung und Praxis. In: H.G. Mehlhorn und K.K. Urban (Eds.), *Hochbegabtenförderung international.* Berlin: VEB Deutscher Verlag der Wissenschaften.

Mönks, F.J. & Span, P. (1985) (Eds.). *Hoogbegaafden in de samenleving.* Nijmegen: Dekker en van de Vegt.

Mönks, F.J., Boxtel, H.W. van, Roelofs, J.J.W. & Sanders, M.P.M. (1986). The Identification of Gifted Children in Secondary Education and a Description of their Situation in Holland. In: K.A. Heller and J.F. Feldhusen (Eds.), *Identifiying and Nurturing the Gifted.* Toronto: Huber.

Mooij, T. (1991) (Ed.), *Onderwijs aan hoogbegaafde kinderen. Muiderberg: Coutinho.*

Nieuwe onderwijsvormen voor 5-13- à 14-jarigen (1965). Rapport van een studiecommissie ingesteld door de Nederlandse Onderwijzersvereniging. Groningen: Wolters.

Overtoom-Corsmit, R. (1991). *Informatieverwerking door begaafde kinderen.* Utrecht: ISOR, Disseration.

Overtoom-Corsmit, R., Dekker, R. & Span, P. (1990). Information processing in intellectually highly gifted children by solving mathematical tasks. *Gifted Education International, 6,* 3, p. 143 151.

Platenkamp, B.R.M. (Ed.) (1991). *Hoogbegaafden in de basisschool.* Den Bosch: KPC.

Renzulli, J.S. & Reis, S.M. (1985). *The schoolwide enrichment model* Mansfield Center, Connecticut: Creative Learning Press.

Révész, G. (1938). De ongedeeldheid der begaafdheidsvormen. *Algemeen Nederlands Tijdschrift voor Wijsbegeerte en Psychologie, 32,* 2.

Révész, G. (1946). *Creatieve begaafdheid.* Den Haag: Servire.

De selectie en ontwikkeling der meer begaafden. (1962). Haarlem: Publicatie van de Stiching Werkcommittee voor Opvoeding tot Democratie.

Span, P. (1983) (Ed.). Themanummer Hoogbegaafdheid. *Pedagogische Studiën, 60,* 5.

Span, P. (1984). Onderzoek naar (hoog)begaafdheid in Nederland. In: *SVO Jaarboek 1984*. Den Haag: SVO Selecta Reeks.

Span, P. & Overtoom-Corsmit, R. (1986). Information processing by intellectually gifted pupils solving mathematical problems. *Educational Studies in Mathematics, 17*, p. 273-295.

Span, P. (1988). Korte historie van het intelligentieonderzoek. In: *Losbladig Onderwijskundig Lexicon*. Alphen aan de Rijn: Samsom, 10 0100-13.

Span, P. (1988). *Begaafdheid, tussen wens en werkelijkheid*. Utrecht: Rede Rijksuniversiteit Utrecht.

Stevens, F.M., Jansen Schoonhoven, A., Kok, W.A.M. & Span, P. (1987). *Onderwijs aan begaafde leerlingen: een voorstel voor beleid*. Utrecht: Vakgroep Onderwijskunde.

Vandenberghe, R. (1964-1965). De begaafde leerling in de Amerikaanse school. *Tijschrift voor Opvoedkunde, 10*, p. 65-106.

Vandenberghe, R. (1965-1966). Bepaling en detectie van de begaafden in de V.S. van Amerika. *Tijdschrift voor Opvoedkunde, 11*, p. 129-169.

IJzerman, T.J., (1970). *Het talentenvraagstuk*. Groningen: Wolters-Noordhoff.

PART 3
Selected Background Papers

Inanç Lycee - A provision for the intellectually gifted children of economically disadvantaged families

Füsun Akarsu
Turkey

The Inanç Foundation, formed by an eminent industrial leader in Turkey, aims to provide secondary schooling (grades 6-11) for intellectually gifted students. The two major characteristics of the school are:

1. it is the first and only school for intellectually gifted children in the country;
2. the target group is the elementary school graduates who come from economically disadvantaged families.

This is to be a privately financed school, but aims at serving wider public interests through the identification and selection of intellectually gifted children from economically disadvantaged families and giving them full scholarships throughout the seven-year programme. It will be a co-educational boarding school located in Gebze, near Istanbul. An area of land of 416,000 square metres has been allocated to the Foundation by the Ministry of Education. The chosen location of the school has the advantages of close proximity to ground transportation and train facilities, as well as being next to the Gebze campus of the Turkish Institute of Scientific Research. Children will be selected from both rural and urban areas from various regions of the country. Student selection includes formal and informal measures combined at various stages. Procedures for the identification of the gifted, and development of the instruments to be used in selection, have already been started.

Although the students are to be selected on the basis of their intellectual potential, the educational process will also develop their artistic, psychomotor and social abilities. Therefore, rather than simply fostering intellectual genius in gifted children, the programme aims for a holistic development of the students. It is expected that a differentiated curriculum will enable students to fulfil their potential in ways and degrees depending on their unique abilities and interests. The ongoing curriculum development studies include co-operation with the Ministry of Education, universities, international organizations and experts. The IB compatible differentiated programmes will help students to have a sound general academic background, as well as high performance in special areas. This is intended to qualify them for admittance to universities of their choice after completing the high school programme.

Another major aspect of the programme is the integration of pastoral care, extra-curricular activites and international relations, with a multi-disciplinary arrangement of content area.

Supported by educational technologies, the school is to be a "learning village" where students, full-time teachers, part-time teaching staff, exchange programme participants, mentors, student's original and host families, researchers and the members of the larger community who co-operate in a variety of activities, will all be co-learners.

A pilot programme was initiated in September 1991, with the first 30 students being accommodated in prefabricated buildings. At present, pre-construction work on the land, architectural project work, identification and selection procedures, curriculum development activities, personnel selection, as well as preparations for an INSET programme, are being carried out by task forces organised by the Foundation. The school is expected to be fully operational by the 1995-1996 school year, with approximately 250 students and about 30 staff members.

Organisation model of the selection, education and development of creative abilities of gifted children in Azerbaijan

Vagif G. Alikperov
Azerbaijan

HISTORICAL BACKGROUND

The educational system in Azerbaijan is a part of the all-union system, the distinguishing feature of which is its rigidity, particularly regarding the uniformity both of teaching procedures and the curriculum content. It is clear that with this education model it is practically impossible to individualize the teaching process to fit the abilities of school children, and the creative elements in the work of the teacher are almost excluded. This situation is worsened by the law of the obligatory secondary education, which practically eliminates the development of motivation in the education process. This is because, regardless of the results actually achieved, all pupils receive the same certificate of the comprehensive secondary education. This situation has given rise to professional apathy on the part of the teachers, the lack of pupils' motivation to study, and an absence of parents' interest in improving their children's education. This is why the level of education of the youth is rapidly falling, and is especially evident in the admittance examinations to higher educational institutions.

Private instruction exists as an alternative to school education. Some good teachers train children from well-to-do families at home for a considerable fee. Usually children are trained for their admittance examinations to higher educational institutions one or two years before they leave school and only in 2-3 subjects. And it is only natural that in this model of education there is no place for the systematic search for and training of gifted children. The result is a lack of competent, functionally trained specialists in science, industry, and agriculture, and such specialists are especially needed now that we are changing over to a market economy.

At the same time, scientist and teachers of Azerbaijan have long recognized the need for a differential education process which would create a favourable educational basis for those children who wish to extend their knowledge in various school subjects. Towards this end different kinds of Residential schools, such as physics-mathematical, chemico-biological, sports, humanities, musical and artistic schools were created in the seventies. Special classes adapted to the interests of children were also introduced in ordinary schools. However, this work was not supported by corresponding scientific research, and most important, the overall model of education remained unchanged or received only slight mechanical modifications, e.g., the amount of the teaching material was simply reduced or enlarged. This is why these schools did not fulfil their aims. In fact, they would not be permitted to, as the Ministry of People's Education of Azerbaijan has no formal right to change the model and the contents of education.

THE CREATION OF "**ISTEDAD**" ASSOCIATION

The overall situation began to change when a group of scientists, teachers and psychologists formed an initiative group for the creation of a non-governmental public organization, the purpose of which was to provide assistance for gifted children, pupils and students to receive a better education and a better development of their intellectual and creative abilities.

The activities of the members of the group started with addresses given on television and radio, articles in the press, and by attending conferences and symposia intended to attract the attention of the government and public opinion to the need to create the conditions necessary for the development and support of children of high ability. With the same goal in mind, the first Summer Creative Camp was set up in the summer of 1989, which brought together about 100 school children of 7-11 grade. A wide variety of lectures and discussions with the participation of well-known writers, poets, scientists, and journalists were organized almost every day.

These activities helped to inform public opinion of the problem. In the winter of 1990, with the support of the government, the Azerbaijan Association "ISTEDAD" was founded. The founders of the Association were three Ministries, the Academy of Sciences and four large social organizations. The main aim of the association is the creation of a new democratic model of education on the national level. This model should provide the necessary conditions for the education and development of individual abilities.

ACTIVITIES OF "**ISTEDAD**" ASSOCIATION.

The following activites are carried out by the Association:

- the search for and selection of gifted and talented children, pupils, and students;
- establishment of special schools, classes, Summer Creative Camps;
- the organization of scientific competitions, publishing of teaching literature;
- assistance to talented young specialists in the realization of their creative abilities in the process of their professional activities.

For the Association to carry out its functions successfully the following steps were first undertaken:

- foundation of the Scientific Research Centre to examine the problems of the development of gifted children;
- opening of the 9 month courses for the training of school psychologists, mostly for rural areas (80% of children in Azerbaijan live in the rural area);
- setting up a computer data bank where the names of all children who were somehow distinguished in different scientific competitions and contests are included;
- foundation of Summer Creative Camps and the Creative School Centre "Samshit".

PROJECTS OF THE "**ISTEDAD**" ASSOCIATION

Nine month courses of school psychologists
(financed by Baku State University)

The main aim of this program is the training of school psychologists from the teachers staff to work with gifted children in each region and city. Annually, about 60 teachers are trained.

Scientific Research Centre to examine the problems of the development of gifted children
(financed by Baku State University)

The main aim of this Center is to study the foreign experience in the identification of, and education and development of children of high ability. In addition, the Center is to adapt existing intelligence tests to take into account the ethno-social specificity of the Republic, and to develop and adapt special teaching programmes and developing courses for talented children.

All these materials are then transferred to local departments of the Association, to Summer Camps, and to the School Creative Centre "Samshit" to be used in the identification of gifted children and in the organization of special classes with more extensive programmes.

Central Computer Data Bank

The main aim of this project is the accumulation of information on gifted children from the whole Republic. The information is collected by local departments, the Ministry of People's Education and by the Executive Council of the Association and is based on the results of scientific competitions. These data are then transferred to the Executive Council and used to extend invitations to these children to attend Summer Camps and the Creative School Centre "Samshit".

Executive Council

The main aim of the Executive Council is the creation of new organization forms for the education and development of creative abilities of gifted children on an individual basis. The Council is responsible for the activities of Summer Creative Camps, "Samshit" school and for annual contests, which are called "Erudite".

A new democratic teaching model for the European Lyceum and a project "Education of gifted children in rural areas" are now being developed. The Council also coordinates the activities of local departments of the Association.

Summer Creative Camps

These camps are in operation for 24 days in July/August and include 50 children. A variety of testing procedures are administered to the children when they attend the camp for the first time. The Camp activities include lessons in standard subjects such as English, computer use, mathematics, physics, chemistry and biology. In addition, a number of other subjects such as "Foundation of Oriental Philosophy", "History of Arts", "History of Religions", "Design", "Ecology", "Politics" are offered to the children.

Creative School Centre "Samshit"

The school organizes lectures by known scientists in different scientific fields, lessons in different subjects according to advanced school programmes, an intensive course in English, computer classes. It also offers a number of extra-curricular courses. Classes are held twice a week for two hours.

Annual "Erudite" Contest

This contest is held in two rounds and is meant for the school children of the 9th, 10th and 11th grades. The tasks of the first round are published in a youth newspaper and are divided in three parts: Physics-mathematics, chemico-biology and humanities. Children who obtain the best results are invited to the second round where they go through intelligence tests. Winners are registered in the Central Computer Bank. Their results and abilities are thoroughly examined and after that they are offered advanced school programmes.

European Lyceum

The aim of the Lyceum is to introduce a democratic education model. This involves the organization of a teaching process which takes into consideration individual abilities of gifted school children. Children will learn two European languages, European history, art and literature. They will be taught in three languages: Azerbaijan, Russian and English. The model of multilingual education is borrowed from the experience of European schools. The Lyceum will be opened in September 1991.

POPULAR SCIENTIFIC MAGAZINE FOR YOUTH "ISTEDAD"

We plan to start publishing this magazine in October 1991. The magazine will contain articles on various scientific problems by the scientists of Azerbaijan, the states of the former USSR and other countries, and will also include different contests and scientific problems.

LOCAL DEPARTMENTS

Local departments are established to organize and coordinate the work with gifted children in each region and city. The main functions are the identification and selection of gifted children, introduction of special advanced courses for school children, organization of town or region contests, exhibitions etc. They will operate in close contact with and have the support of local bodies of people's education and municipal bodies.

PERSPECTIVES FOR FUTURE ACTIVITIES

The structure of the Association will change with time and new forms and directions of activities will appear. Among other things, we are considering the creation of:

- international Summer Creative Camps;
- new educational programmes in cooperation with European researchers;
- local problems for the work with gifted children in every region and town;
- exchange programs for school children, students and teachers;
- organization of joint contests, competitions, exhibitions, etc.

In the sphere of psychology of the gifted the following research is planned:

1. Psychological provision of the identification of giftedness. This research focuses on the problem of early identification of general and special gifts, the creation of a search system, and the selection of gifted persons.
2. Adaptation and development of identification instruments of general and special gifts.

In conclusion, I would like to offer the cooperation of the "ISTEDAD" Association in the realization of different international projects. We are ready to cooperate with all social organizations in the sphere of development and support for talented young people.

Research on the gifted in Italy

Ornella Andreani Dentici[1]
Italy

CULTURAL AND HISTORICAL BACKGROUND

Attention for the problems of individuals above-average either in general ability or specific talents has fluctuated historically, which can be attributed to the influence of social structures and ideologies in different periods. Intellectual superiority was viewed favorably in periods like the Italian Renaissance, the Protestant Reformation and the Industrial Revolution, when the ideology of intellectual élites conformed to the values of a dominant class to which it offered a theoretical justification of privilege and the instruments to perpetuate it. In fact, when people accepted the principle that aristocrats were superior by divine right, it was natural to reserve a privileged education to their children. Later, when economical power substituted for divine right, excellence became a characteristic of capitalistic and advanced societies, finding theoretical support in the protestant ethic, which viewed success as a sign of Grace, as well as in the evolutionist theory of natural selection, which offered a scientific basis for the dominance of the biologically and intellectually gifted.

This complex of causes explains the development and flourishing of research on the gifted in the Anglo-Saxon countries, where there has been a passage from the study of genetic differences to the field of applied psychology, permeated by the philosophy of "the right man in the right place" and by the hope that a democratic society can develop optimally all individuals' capacities and talents. Initially, a series of educational experiments were developed, such as the formation of different classes or groups for the gifted, school acceleration, and enrichment. After the launching of the first Sputnik the interest for educational problems

1 Prof. Dr. Ornella Andreani Dentici, Università degli Studi di Pavia, Istituto di Psicologia dell'Università, Piazza Botta 6, 27100 Pavia, Italy

received a stronger impulse. Paradoxically, the communist countries, where the school had a very centralized organization and was strictly connected with political planning, concentrated their efforts on programs to identify and foster individual abilities in scientific and technical fields, as well as in sport, music and dance. Providing the best education for the gifted was a means to attain the superiority in international competition which was a social-political duty in communist and socialist countries. The rich European democracies tended to avoid élitism and selection based on talent in the name of equality of opportunities.

It is interesting to note that the problem of the identification and curricular help for the gifted has been rapidly transforming in studies on creativity, directing the attention to the differences between divergent and convergent thinking and to the relationship with personality traits and dynamic processes. The sixties and the seventies shifted the focus of research to environmental factors, and the problems of cultural disadvantage, social classes and minority groups became dominant. This shift fostered many changes in the concept of intelligence, brought into question the validity of traditional methods of measurement, and stimulated a search for new ways of fostering and nurturing thinking abilities.

RESEARCH IN ITALY (1960-1980)

The same trend can be seen in Italy. In 1962 a program for the identification and assistance of gifted was started (IARD Association in Milan), sponsored by Brambilla, Visalberghi and Passow. After the first phases of research (Calvi, 1965; Andreani & Orio, 1972) the focus shifted to enrichment programs open to all pupils (De Benedetti, 1972), then from the gifted to the causes of under-achievement (Andreani et al., 1974) and later to youth problems in general (Cavalli & De Lillo, 1988).

During the last twenty years in Italy there have not been important research projects on giftedness, although there have been some studies on creative thinking (Calvi, 1972; Pagnin & Vergine, 1975; 1977). Many others have examined logical thinking from an experimental or developmental point of view, but without special attention to the gifted. In the same period in Europe, particularly in Great Britain, Germany, Netherlands, Hungary and Poland, we assisted in longitudinal studies, in programs of educational intervention and in the foundation of the European Council of High Ability which tries to develop communication between academic research and educational policy.

Up to now in Italy there has been a dissociation between official educational policy and the private sectors. Public schools claim to adhere to an equalitarian ideology, and the only initiatives for special groups are in favour of the handicapped. At the same time private associations provided grants and prizes to gifted pupils, have organized national competitions for mathematics, scientific inventions, pictures, film production. There are also some university colleges, like Ghislieri and Borromeo in Pavia, and Scuola Normale in Pisa, which admit only selected students and promote their development through many courses parallel to regular university lectures. In the Catholic field there has been a unique initiative promoted by a priest in Sicily, the so called "Villaggio per super dotati" (Village for the Gifted) but it started and continued without any psychological counseling, more in the form of assistance to disadvantaged then as a systematic educational program.

The program undertaken in the seventies by the Identification Assistance and Research on Gifted Children Association (IARD) in Milan developed through the following phases:

1) construction and standardization of tests for the 11-14 year level (one test which measures the g-factor and two tests of verbal and mathematical ability) (Andreani & Calvi; 1972);

2) identification of the subjects above the 90th percentile in two of the three group tests, representing upper, lower and middle classes;

3) intensive study of a sample of 128 gifted 11-14 year old children using an intelligence test, projective techniques, an interview, personality inventories, teachers' and parent's assessment (Andreani & Orio, 1972);

4) program of grants and social assistance for the children; special educational programs (extra-curricular "enrichment activities") open also to other children in the school;

5) research on creative thinking (Calvi, 1972a,b) research on social, family and personal factors in school achievement (Andreani et al., 1974);

7) extension of enrichment programs to large samples of normal children in ordinary schools;

8) guidance and vocational programs on a large scale.

The following conclusions emerge from this research:

1. The use of a battery of good traditional tests has enabled us to choose a sample which has confirmed a high level of intelligence in WISC.

2. The sample also included a good group of creative subjects, although above 125 IQ there was no correlation between intelligence and creativity indexes.

3. If we want to identify potentially creative subjects in a large population, the best strategy seems a *combination* of methods to be used in *sequential phases*: first, the application of group intelligence tests, followed by the application of more sophisticated tests in *individual form* to subjects of the highest percentiles. This second phase should include tests of creative thinking, and other indicators of creative potential from projective techniques and from sampling of spontaneous productions (like writing, drawings, art performances), assessments of interests, hobbies, motives and aspirations.

4. Our results confirm the observations made by many others: the gifted are the product of an interaction between genetic and environmental factors; they are frequently the first born in small families, of good cultural level, have been stimulated by young parents, and are highly achievement motivated. Nevertheless, they are more independent of the family and many of them show signs of detachment and criticism towards parents and teachers.

   The intellectual profile is of a harmonious integration of all mental functions, illustrated in perceptual articulation, in speed and precision, in logical, verbal and mathematical thinking, in fluency and flexibility of thinking, high creative potential. They also have a wide variety of interests.

5. Personality, as studied by Rorschach projective tests and interviews, shows a more precocious growth rate, with a pattern more similar to adolescents and adults than to peers; there are signs of imagination and human empathy, rich inner life with signs of stronger drives for aggression and

sex, usually well controlled; in general there is good adjustment, although some notes of anxiety, excess of intellectualization and many signs of low conformity might lead the gifted to oppose adults and social norms.

6. Although the overall picture is that of bright, normal subjects with a well integrated personality, there is a danger that this picture modifies in the "*gifted-syndrome*", which emphasizes intellectual traits and inner life, and leads to excessive abstraction and perhaps to autistic thinking, or exaggerates emotional reactivity leading to anxiety or impulsive behavior.

Variety and richness of interests itself may evolve in a negative direction transforming itself in excessive dispersion and lack of concentration. Even the favourable conditions of small families may transform into stress factors when the children are confronted with excessively high expectations, particularly from parents who try to compensate for their own frustrated ambitions. We found only a few such cases in our sample, but we suspect that they become more frequent during adolescence, when intellectual superiority may easily bring the gifted to isolate themselves from peers, to clash with social norms, and to revolt against adults.

EDUCATIONAL PROBLEMS

An examination of school performance of our gifted sample revealed approximately one third bright students, one third above-average, and one third who failed scholastically. The latter group shows two types of maladjustment: if they are highly intelligent they show conflict between an excessive need for achievement, which leads to obsessive perfectionism and anxiety, and a need for autonomy which induces them to drop out of school. If they are highly creative they present low control of drives, difficulty in identification with parental models, low frustration tolerance, and intolerance of conventional teaching methods.

The dilemma involves avoiding both the danger of an élitist education and the danger of wasting high intellectual potential and talents in a gray, monotonous school routine. The educational choice to be made must not only meet the psychological needs of the gifted but at the same time create an equality of opportunities for all.

If we want to satisfy the needs of exceptional children it is not necessary to segregate them into special groups, risking isolation from peers, intellectual detachment from reality, narcissistic self-consciousness and excessive pursuit of personal success. We should experiment with more efficient forms of individualization. Teaching should offer different levels of difficulty in the same class, but the teacher should not encourage competition, but appreciate the performance of less clever pupils.The need for achievement of the gifted should be *task-oriented* more than *self-oriented*, the goals should be chosen at individual level of difficulty, perhaps emphasizing social relevance. They should be taught to discuss the social implications of science and technology and to exercise moral responsibility in their choices. It is necessary to carry out a more systematic analysis of the relationship between moral and cognitive development, and to pay more attention to the emotional problems of children and to the learning of social behavior. The typical difficulties of adolescence, e.g., role adoption in adult society, should be faced with new educational strategies which develop autonomy, responsibility, self-esteem and confidence. The gifted should have optional activities to develop specific talents, but should maintain contact with peers in other activities.

RESEARCH ON MORAL JUDGMENT (1989-92)

When I again took up research on giftedness, and reviewed recent work, I was struck by the amount of research concerning cognitive factors and the scarcity of work on affective and social factors.The latter were usually studied with regard to emotional problems connected with parental attitudes, excessive ambition, socio-economic disadvantages or with emotional reactions to a rigid school atmosphere.

The area of social competence was infrequently treated, and was restricted to topics like social relationships, communication, or leadership, and neglected the important area of moral reasoning, that should be very relevant for a complete and well integrated development of personality. In addition, the study of the development of social and moral reasoning, especially during adolescence, might provide a more concrete basis for examining logical reasoning, and might suggest ways to prevent deviant behavior. Finally, I believe it is essential that intelligent people use their superior capacity to comprehend, interpret and anticipate the needs of others and to learn pro-social behavior. To examine these issues, we started a research project on moral reasoning in gifted adolescents. The first results were presented at the ECHA Conference in Budapest and at 9th World Conference on Gifted and Talented Children, in The Hague in 1991 (see Mönks et al., 1992).

We examined the responses to six moral dilemmas, comparing subjects of low, average and high ability in a sample of 1141 students attending the last year of secondary school (age 17-19) in three towns of North, Center and South Italy. Two groups of very high (5 % of the total sample) and extremely high students (2%) were individually interviewed in order to explore the motivation of their choices. As expected, the gifted gave more responses at the formal level (greater choice of items concerning general and abstract principles, greater support for law and social contracts, more coherence). In the average and lowest groups we found a more sentimental, humanitarian orientation, with a sympathetic attitude towards other people's feelings, which constituted the basis for a more intuitive morality (Andreani & Pagnin, 1991).

On the basis of these results, we proposed a two-level model of moral judgment: a morality of sentiments (based on common sense and sympathetic feelings) and a rational morality (centred on rational principles and the attempt to weigh the consequences of actions). The first one was more frequently observed in females and in average-low intelligence group, the second one in males and in gifted subjects. We suggested that the moral reasoning of the gifted is not necessarily superior, since the greater logical coherence and the high *need* for *achievement* may neglect the immediate feelings of empathy and the moral inhibitions. Both levels - the more formal and the more intuitive one - can lead to moral behavior and may be present in the same subject. The educational challenge is to reconcile logical coherence and general application of law with an empathetic attitude.

In the second study a more analytic examination of the relationship between the level of intelligence, formal reasoning, creativity and moral judgment was made (Andreani & Pagnin, 1992, in press). New dilemmas concerning war and peace were added and analyzed following the Rest model for the analysis of three components: interpreting the situation (comprehension and role taking), determining the ideal course of action (reference to norms and principles), and motives and values used in the decision (scored by open questions and a value inventory). The hypothesis was that highly gifted, as defined in reasoning tests, would manifest a better definition of principles, and would be more coherent in the reasoning behind their choice. The highly gifted, as measured in creativity tests, would attend more to complexity of situations, would demonstrate more role-taking, and provide more original solutions. To test this hypothesis we compared the response to moral dilemmas in high, average and low intelligence and creativity groups in a sample of 371 high school students (age 15-17). A smaller representative group (N=32) were individually interviewed to determine the motives behind each choice and the values of the subjects.

The hypothesis of a positive relationship between creativity tests and the scores in moral reasoning was not confirmed: The correlations between the comprehension of situations, role taking and stage of formal reasoning were not significant. Although the subjects higher both in intelligence and creativity tests showed a higher orientation to general principles and to the analysis of consequences of actions, there was a difference between subjects in specific choices made, accounted for by differences in personal values, principles and ideologies.

As a group, gifted subjects (highly intelligent and/or highly creative) appear more independent, self-confident, ready to express sentiments and attitudes even if unconventional. However, the principles they defend can be very abstract and the coherence of formal reasoning concerns the *form* more than *content* of moral acts (see Bandura, 1991). The two kinds of moral judgment ("sentiment" moral and "rational" moral) which we found in the first study were both present in this gifted group, but appear as two components and not as two stages of the process of moral reasoning. The sentiment morality adheres to common standards, an immediate comprehension of others feelings, an empathetic attitude, while the rational morality requires a calculation of benefits of the different actors, a forecast of the outcome of actions, and a balance of the different aspects of rights.

The educational process should perhaps integrate both components, where science education develops logical deduction from principles, while the humanities, through the analysis of novels and poems, and through philosophy and history, increases the capacity to identify with others. Moral education projects could involve discussion of real life cases and dilemmas and cooperation projects. Although the gifted have a more precocious intellectual growth, they cannot anticipate the experiences of life or of other people. They need to reflect on the contradictions which can occur between principles and values, which can serve as an impulse to their development. In our opinion the most important problem is not that of a precocious identification of gifted, but of an education which enables each person to reach the best possible level for him and to use his abilities in the service of the community.

References

Andreani, O. & Orio, S. (1972). *Le radici psicologiche del talento.* Bologna: Il Mulino.

Andreani, O. et al. (1974). *Classe sociale, intelligenza e personalità.* Bologna: Il Mulino.

Andreani, O. & Pagnin, A. (1991). Moral reasoning in gifted adolescents: Cognitive level and social values, *European Journal for High Ability, 2,* in press.

Andreani, O. & Pagnin, A. (1992). Moral judgment in creative and talented adolescent, *Creativity Research Journal,* in press.

Bandura, A. (1991). *Social cognitive theory of moral thought and action.*

Calvi, G. (1965). *Ricerche sulla creatività.* "Contributi dell'Istituto di Psicologia dell'Università Cattolica", XXVVII. Milano: Vita e Pensiero.

Calvi, G. (1972). *La misura dell'intelligenza.* Bologna: Il Mulino.

Calvi, G. (1972). *La creatività.* "Nuove questioni di Psicologia". Brescia: La Scuola.

Cavalli, A. & De Lillo, A. (1988). *Giovani anni 80: secondo rapporto lard sulla condizione giovanile in Italia.* Bologna: Il Mulino.

De Benedetti, M. (1972). *La scuola media impossibile.* Bologna: Il Mulino.

Pagnin, A. & Vergine, S. (1975). *Il pensiero creativo.* Firenze: La nuova Italia.

Pagnin, A. & Vergine, S. (1977). *La personalità creativa.* Firenze: La nuova Italia.

Gifted girls: A case of wasted talent?

Rosemary Ayles
United Kingdom

A number of studies on gender differences in education and employment indicate that the particular needs of gifted and talented girls are not well recognised, and their abilities are underpromoted to a greater extent than in the case of boys.

There is ample evidence from work on gender differences in education that:

a) most teachers have a lower expectation of girls than of boys;
b) girls have a lower opinion of their own abilities than do boys (Chandler, 1980; Deem, 1984; Mahony, 1985).

Judging from research conducted in several countries, it also appears to be the case that, in all school subjects and with both male and female teachers,

a) boys receive more instruction, praise and criticism, and
b) girls are just as likely to volunteer answers, but boys are more likely to be chosen to speak (Kelly, 1988).

Girls tend to be more vague about career aspirations and to aim lower than boys of similar ability (Kerr, 1983). Yet an EC action research project found that "in the case of young girls in particular, if they are not clear about their aspirations and their aptitudes then despite all their efforts they tend to underachieve" (Valabreque, 1989). In mixed school particularly, there is considerable social pressure on adolescent girls to underachieve (Chandler, 1980; Mahony, 1985). Choice of subjects in adolescence continues to reflect gender stereotypes, with boys tending to opt for technology and the sciences and girls for the arts (HMI, 1989; Pickersgill, 1989; Jones, 1989).

A report of the European unit of Eurydice points out that throughout Europe "girls are grossly under-represented in sectors relating to the exact sciences and new technologies" while "those courses mostly preparing for teaching, for occupations related to information, documentation, and for the services and medical and paramedical professions have a majority of female students" (Barriuso-de Ceuster and Moisan, 1990). Yet in 1990 although only 29% of the GCSE entry of Physics was female, girls achieve 56% of the top grades (A-C) (1990 GCSE Inter-Group Statistics).

These findings relate to all pupils, but recent work suggests that there are particular implications for very able girls, and that the findings of Chandler (1980) and Mahony (1985) cited above, apply particularly to this group (Eyre, 1990; Ayles and Fuller, 1991). Other studies concerned specifically with very able girls indicate that, in addition to the pattern outlined above, highly able adolescent girls in mixed schools resent anything which sets them apart from their peers and will often, to a greater extent than boys, deliberately underperform in order to achieve "normality" (Wallace, 1983; Ayles, 1991).

This pattern of unjustifiably low expectations, lack of confidence, social pressure to underachieve and a bias away from science and technology, established during the first years of education in school, is reflected in the higher stages of education. Women students in the UK are still under-represented in universities and polytechnics, although they are well represented in lower level courses of further education, as the following table illustrates.

Students in Higher and Further Education for the Academic Year 1987/1988

	Universities and Polytechnics	***Further Education***
Men Home Students (full time)	***337,000***	***69,000***
Men Home Students (part time)	***219,000***	***891,000***
Women Home Students (full time)	***272,000***	***74,000***
Women Home Students (part time)	***140,000***	***1,863,000***

In every Member State of the EC the percentage of women in Higher Education increased during the 1975-1985 period, yet only in France and Portugal did it reach 50% or more. In three Member States it remained below 44%. At the level of postgraduate education the percentage of women was considerably lower (Barriuso-de Ceuster and Moison, 1990).

Recent figures on women in the professions show an even more marked pattern of under-representation. For example, in the UK in 1990 only 0.5% of engineers, 3.2% of surgeons, 8% of architects and 9.7% of chartered accountants were women (Packard, 1990). Furthermore, figures published by the UK Association of University Teachers (1990) indicate that women account for 7% of senior managers in industry, 3% of University Professors and 2% of University Vice-chancellors.

Under-representation of women in senior positions is widespread. Since there is no evidence to suggest any corresponding imbalance in the numbers of highly able and talented boys and girls in schools, one has to look elsewhere for explanations. It would seem reasonable to suggest that, at least in part, the pattern is set in our schools.

The Commission of the European Communities, through its Resolution in Equal Opportunities for Girls and Boys in Education (85/C 166/01), the subsequent action programme (Euryclee 89 Info 4), the establishment of the IRIS Network (1988) and NOW (1990) has raised awareness and improved the position of girls and women in education and work. However, these are measures designed to help all girls and women, and as such tend perhaps to address the needs of the majority. There is little mention of high ability and it is impossible to know the extent to which these measures have affected the exceptionally able and talented.

In conclusion, it appears to be the case that unless the effects of gender differences on education are taken into account, measures to improve education for the most able will, by failing to recognise the implications, perpetuate the inequities created by attitudes and practices which foster gender inequality. More research is needed if 50% of gifted children and adolescents are not to be kept in a disadvantaged position. Not only is it vital to foster high ability in children, but perhaps particularly so in girls since it may be less readily recognised. This, in turn, points to a need to raise the awareness of both teachers and parents of the importance of identifying high ability in young children and, in particular, of recognising the pressures which can militate against high levels of achievement in gifted girls of all ages.

References

Ayles, R. (1991). *Mixed 6th Form: Hidden agendas.* University of Reading, in preparation.

Ayles, R. & Fuller, M. (1991). *TVEI in ITT Report.* University of Reading, in preparation.

Ball, C. (1990). *More means different: Widening access to higher education.* Royal Society of Arts, London.

Barriuso-De Ceuster, A. & Moisan, C. (1990). *Girls and boys in secondary and higher education.* Brussels: European Unit of Eurydice.

Chandler, E.M. (1988). *Education Adolescent Girls.* George, Allen & Unwin.

Deem, R. (Ed). (1984). *Coeducation Reconsidered.* Open University Press.

Eyre, D. (1990). Does an active policy of enrichment and extension for more able pupils have an effect on gender differences? Unpublished MEd Thesis, University of Reading.

HMI (1989). *Education Observed 14: Girls learning mathematics.* London: Department of Education and Science.

IRIS (1990a). *Annual Report 1989.* Equal Opportunities Unit.

IRIS (1990b). *Bulletin No. 8.* Equal Opportunities Unit.

Jones, L. (1989). Context, confidence and the able girl. *Educational Research, 31,* No 3, p. 189-194.

Kelly, A. (1988). Gender differences in teacher-pupil interactions: a meta-analytic view. *Research in Education, 39,* p. 1-23.

Kerr, B. (1983). Raising the career aspirations of gifted girls. *Vocational Guidance Quarterly.*

Mahony, P. (1985). *Schools for the Boys?* London: Hutchinson.

Packard, L. (1990). *Women in the Profession.* London: Law Society.

Pickersgill, D. (1989). Girls and Science. *Education in Science.*

Valabreque, C. (1989). Research - Action: Girls and New Technology, Report of the First Phase. *Euryclee Info 4.*

Wallace, B. (1983). *Teaching the very able child.* London: Ward Lock.

Gifted children: Research and education in Spain

Javier Berché Cruz & Leopoldo Carreras Truño
Spain

The first studies in Spain which examined the education of the intellectually precocious, gifted and talented children, were begun around 1936 at a school in the city of Barcelona, whose name was "escuela del Bosch". Unfortunately, within a few months of the start of this project, civil war broke out, and all documentation related to this research was lost.

No other work in Spain on this topic appears until 1947, when the "Instituto de Selección Escolar" (Scholastic Selection Institute) was founded in Vizcaya. The institute offered scholarships, whose name was (and still is), "Juan Manuel Alonso Allende-Fidela Aregui", in order to reward those students with a high academic development. It also started a school where those students received individual education. Many of these students now occupy important positions in the cultural life of Spain. The school had to be closed in the early sixties due to a lack of funds, but the scholarships are still awarded and there are plans to re-open this school. No further attention was given to the topic until 1981, when Prof. Garcia Yagüe published a study on gifted children, done at the Department of Psychology of the University of Madrid.

At the beginning of 1987 the Association for the Development of Creativity and Talent (CREDEYTA) was founded by a group of parents of gifted and talented children. The purpose of this Association was to direct attention to the problems of gifted children, whose interests and special problems were inadequately treated in the Educational Laws of Spain at that time. Since then, the work of the Association has aimed at mobilizing public opinion and interest in the problems of giftedness. Some of the Association's accomplishments are:

in the Educational Laws of Spain at that time. Since then, the work of the Association has aimed at mobilizing public opinion and interest in the problems of giftedness. Some of the Association's accomplishments are:

A. Providing better information about the real nature of giftedness and de-mythifying the public image of the talented. During the past four years CREDEYTA's team of education councellors has studied about 800 gifted children from all regions of Spain, providing help for an educational need that few people realized had existed. The result has been a growing interest in giftedness in a wide variety of fields related to education. As a consequence, the further development of CREDEYTA has moved from theory to more practical concerns. It has now alligned with the Instituto de Selección Escolar, and will soon form a similar alliance with a new Association founded in Valencia whose name is AVAST.

B. Research activities have focussed on the improvement of the identification and diagnosis of giftedness. In this area the group has done:

1. A study on the Silverman-Rogers Parents' questionnaire. This was a two year project that adapted and slightly modified the questionnaire. Results of the study were presented at the first ECHA Congress in Zurich in 1988.
2. A study on the identification of mathematical talent. The first phase of the study adapted and modified the mathematical sections of the Scholastic Aptitude Test of the Council Board. The second phase will involve testing in 20 schools of the city of Barcelona to identify the superior students in mathematics around the age of 12 or 13 years old. This project is similar to the S.M.P.Y. Program (Study of Mathematically Precocious Youth) carried out in several states of the U.S.A. and headed by Prof. Julian C. Stanley.

C. The educational activities of CREDEYTA can be summarized as follows:

PEDAGOGY

1. Individualized attention for those children who need reinforcement in deficit areas as well as to promote the development of creativity in other areas. This individualized attention takes the form of tutorials.

2. Extended Curricular Program for talented children in Primary Education. This Program is being carried out in three schools around Barcelona. The children in this program are involved in curricular enlargement and creative potential activities for 3-4 hours a week outside of their usual classrooms. The first year of the program involved a test period of two months, to develop some experience with it. Fuller implementation has begun this year, although it can not be completely developed because of a lack of necessary funds.

PSYCHOLOGY

Emotional reinforcement is the key to the development of gifted children, which is why efforts are made to integrate them into society. Children receive continuous guidance both in terms of preventative as well as therapeutic interventions.

ADVICE TO PARENTS

It is obvious that continuous advice to the parents of gifted children is essential for the psychological development of the child.

All levels of society have to be conscious of the need for proper education for the gifted child. To develop this consciousness, it has become necessary to give scientific credibility to the actions undertaken for the benefit of the gifted. To meet this end in Spain CREDEYTA has been involved in:

- 1st. International Congress of EUROTALENT, Barcelona 28/31 August, 1989.
- 1st. Symposium on the education of the gifted child, Barcelona, 4/6 May, 1990.
- For the 1991-1992 school year, 3 symposia will be organized for teachers, educationalists, psychologists, pediatricians, and parents.
- Publications in scientific magazines.
- Attending conferences on the topic of giftedness.

To sum up, this professional team has been and continues to be active in promoting better understanding of and education for the gifted in Spain. Better economic assistance would improve the quality and quantity of its' work.

Mathematical giftedness - Some questions to be answered

Vladimir Burjan

Czechoslovakia[1]

In this paper I wish to identify and briefly discuss the main questions and problems that should be examined when studying the problem of mathematical giftedness. These questions and problems are:

1. What is mathematical giftedness (MG)?
2. Different types of MG
3. Why to identify mathematically gifted pupils (MGPs)?
4. When (at what age) should MGPs be identified?
5. How to identify MGPs?
6. Which assessment procedures (which types of tasks) are suitable to identify MGPs and which are not?
7. How should MGPs be fostered?
8. Social and psychological aspects of MG
9. Teachers of MGPs

Most of these problems have been discussed in many scientific papers. However, it does not seem to be an easy task to arrive at clear conclusions. The following are some suggestions for further discussions and research in this field.

WHAT IS MATHEMATICAL GIFTEDNESS?

It is by no means easy to answer this "crucial" question. Different answers would probably be given by pure mathematicians (scientists), by mathematics teachers,

---

[1] This contribution is based on the author's draft report of a working group at a conference on "Assessment in Mathematics Education and its Effects". The conference, held in Calonge (Spain) in April 1991, had been organized by the International Commission on Mathematical Instruction (ICMI).

and by psychologists. However, the experience of teachers working daily with mathematically gifted pupils (e.g. in special schools and classes) suggests that the importance of theoretical "definitions" of MG should not be under-estimated. It might be useful for mathematics teachers to be aware of the differences between the following three groups of pupils:

- high motivation (= showing interest in mathematics)
- high performance (= good results in achievement tests)
- mathematically gifted (= ???)

The following diagram gives an illustration of what seems to be the situation with our students:

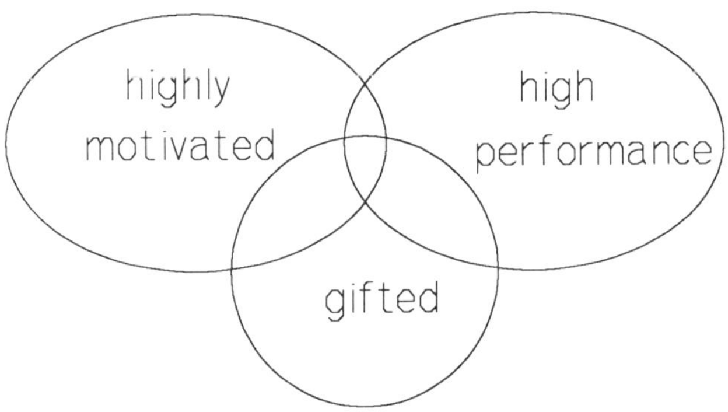

Among many conclusions that can be inferred from this diagram one is of special importance:

The gifted pupil may not necessarily demonstrate high performance (especially when standard assessment methods are used) and may not be highly motivated (before being identified and fostered).

DIFFERENT TYPES OF MATHEMATICAL GIFTEDNESS

The experience of several teachers engaged in systematic work with MGPs suggests that there are several types of mathematical giftedness. In fact we expect a wider variation in the forms and manifestations of giftedness than we see in ordinary students. The main two "archetypes" could be characterized as follows:

The problem-solver type

- is extremely skillful (successful) in solving both standard and non-standard (challenging, olympiad-type) problems
- is able to perform optimally in timed tests and competitions (is often a successful olympiad contestant)
- his main interest is in searching for solutions to problems that have been posed by somebody else.

The investigative type

- is interested (and successful) in detailed investigation of open-ended problems
- may be slow and not able to perform well in a definite time and under time-pressure, prefers a long-term investigation of some problem
- his/her main interest is in asking (and answering) own questions, introducing own concepts, inventing own methods or algorithms.

A more detailed analysis of the differences of these two archetypes can be found in Burjan & Vrba (1990). The question regarding different types of giftedness should be examined because different types of giftedness would have to be both identified and fostered in different ways.

WHY IDENTIFY MATHEMATICALLY GIFTED PUPILS?

We wish to identify giftedness because the special talents of the mathematically gifted are a particular treasure for society. Since there may be few good teachers for the gifted, and because, we assume, they will learn well from one another, special schools and classes for gifted are appropriate.

WHEN (AT WHAT AGE) SHOULD GIFTED PUPILS BE IDENTIFIED?

We must identify the gifted at the optimum age – not so early that they become bored with special study and not so late that we miss the best opportunity to nurture their talents.

HOW TO IDENTIFY GIFTEDNESS?

This is one of the crucial problems. Identification of the gifted is often a multi-stage process. The initial stage is usually the individual teacher's observation, either through regular work with the students or through applying assessment procedures designed for all the students. In this context two important problems arise:

1. Are the teachers being prepared sufficiently for the identification of gifted pupils in their pre- and in-service training? Is it possible to advise the teachers on how to efficiently identify a MGP?
2. Are the widely used traditional assessment procedures suitable for identifying a gifted pupil (e.g. for distinguishing him from the highly attaining but not gifted one?)

In several countries (e.g. Czechoslovakia) special procedures for identification of MGPs are being designed and constantly evaluated and improved (for more details see Burjan, 1991). The designers of such procedures have to face many problems, e.g.:

- since there are several types of MG, we may need different forms of identification and assessment procedures
- we must be flexible with students, as their responses to tests may hide their real gifts.

In several countries (e.g. Russia) Mathematical Olympiads and similar competitions work well in identifying gifted students. However, evidence from Czechoslovakia shows that other gifted students (e.g. the investigative types described above) may be missed if this is the only approach used. In Hungary, teachers and parents are urged to contact specialists if they suspect their students are truly gifted. The specialists can then screen and, if indicated, help these students.

WHICH ASSESSMENT PROCEDURES (WHICH TYPES OF TASKS) ARE SUITABLE FOR IDENTIFYING GIFTEDNESS AND WHICH ARE NOT?

It is difficult to formulate general rules for evaluating mathematical tasks according to their appropriateness for identification of MGPs. However, some recommendations seem to be generally acceptable. Thus, for the purpose of identifying MGPs it is recommended to use:

- open-ended investigations and open-response problems rather than multiple-choice
- problems allowing several different approaches
- non-standard tasks rather than standard ones
- tasks focusing on high-order-abilities rather than on lower-level-skills
- complex tasks (requiring the use of several "pieces of mathematical knowledge" from different topics) rather than specific ones (based on one particular fact or technique)
- knowledge-independent tasks rather than knowledge-based ones.

The best way to answer this question seems to be by presenting samples of concrete problems that have been used and that have proved to be efficient for the purpose of identification of MGPs. (More than 20 such problems can be found in Burjan, 1991).

The problems of the development of gifted pupils and the social and psychological aspects of mathematical giftedness are beyond the immediate scope of the present summary. Nevertheless, such problems cannot be ignored when attempting to grasp the problem of MGPs in its entirety. (For more information about fostering of MGPs in Czecho-Slovakia see Burjan, 1990).

SOME SUGGESTIONS FOR FURTHER RESEARCH IN THIS FIELD

The present situation seems to be like this: The Middle and Eastern European countries (e.g. Soviet Union, Hungary, Czechoslovakia, Poland) have gathered much experience with the identification and development of mathematically gifted children which might be interesting for a wider audience. Unfortunately, these countries are currently in a difficult economic situation which makes further research in this field impossible. On the other hand, some of the developed Western countries might be interested in the problem of gifted pupils and they possess some financial resources. Thus, we suggest that some international (or maybe bilateral) projects be started which would be mainly financed from the West (at least in its initial stages) and for which the "know-how" would come from Eastern and Middle European countries with much experience in this field. Such projects and studies might prove fruitful for both sides.

There are many special problems in the area of giftedness. In particular we urge further studies in the following:

1. A survey of the practices of identification, fostering and assessment of MGPs in countries around the world.
2. A compilation of case studies as a source of direction for others.
3. A typography of different types of mathematical giftedness.
4. Samples of problems which proved appropriate for identifying various kinds of mathematical giftedness.
5. Social and psychological aspects of mathematical giftedness.
6. Problems of teachers of MGPs.

References

Burjan, V.(1990). Fostering of mathematically gifted pupils in Czechoslavakia (a survey). *Acta Mathematica Didactica.* Bratislava: Comenius University (in English).

Burjan V. & Vrba, A.(1990). The Role of Mathematical Compctitions in Popularization of Mathematics in Czecho-Slovaka. In A.G. Howson and J.P. Kahane (Eds.). *The Popularization of Mathematics.* ICMI Study Series. Cambridge: Cambridge University Press.

Burjan, V.(1991). Identification and assessment of mathematically gifted pupils. ICMI Conference "Assessment in mathematics education", Calonge (Spain), 11-16.4.1991.

The educational system for gifted and talented people in Romania

Carmen Cretu
Romania

There has always been an interest in gifted and talented children and adolescents in Romania, which shows the general understanding of the special role that such personalities play in maintaining the cultural profile of the nation. Unfortunately, the degree of interest has not been constant nor correlated with policy in areas of school legislation, provision in school, social and professional promotion, or research.

In early 20th century, the Romanian educational system was still influenced by an 1864 law which had provided a unitary character to the pre-academic training in Moldavia and Wallachia, and had organized the academic training, founded at Iasi in 1860 and at Bucharest in 1863. In 1904 the then Minister of Education, Spiru Haret, labeled this law as "anachronic and one-sided" and suggested a complete reform "not only in its external aspects but also in its inner nature, in spirit, tendencies and ideal". The Report addressed to His Majesty King Carol concerning the activity of the Ministry of Public Instruction and Cults contains *the first input* for the special education of gifted children and adolescents.

The primary school was organized in terms of specific categories in the rural and urban environment. There were "regional holidays" for children, keeping with the school requirements, and also with the family life in various areas. The primary school curricula comprised complementary courses and various ways and means of extra-school training, according to the particular scope.

High schools were organized as an eight-year programme (gymnasium and Lyceum). The first four forms involved an obligatory curriculum for all pupils.

In the fifth form there was a division into: a) classical, b) real, c) modern studies, based on students' abilities. If initial student placing proved to be incorrect, the law allowed shifting to another form of education based on some extra exams. Professional training was organised for the fields of agriculture, crafts, trade and home service (for girls).

An interesting strategy for identifying gifted pupils was adopted as an educational policy, i.e., a special exam at the end of the high school program, provided by this law (art. 19). This exam was not the equivalent for the well-known Baccalaureat (A-levels). The assessment methodology for the candidates was described in art. 96: "The examining commission will assess the assimilation of the objects studied and the maturity level of thinking, not the volume of knowledge. The Commission will see whether the pupils can arrange the data, choose what is essential, give up useless details, unfold ideas clearly, and give a systematic presentation within the time limits at their disposal. The Commission will evaluate the originaltiy of the paper, the style, the fluidity and the flexibility of the ideas". The specific character of this examination consisted in the freedom it offered the candidates to demonstrate their preferences and abilities, which was next to impossible in the baccalaureat. Many children, not outstanding in virtue of their previous results, had a good opportunity to do well, and often did so, in this examination.

The high mental ability students from Iasi and Bucharest Universities benefited by facilities and fellowships, irrespective of their family income (such stipends were provided by Caisse des Ecoles, Fundatia Carol, Ministry of Public Instruction, Fondurile Hillel, Adamaki etc.).

The spirit of the Reform was marked by the importance that S. Haret gave to the individual's *ethical quality*. He went so far as to consider that intellectual output depended on ethical quality. The making of boarding schools emerged from the belief that "the people's vigour is to be found in villages" and "the ethical side of peasants is more marked than townsfolk's". The boarding high schools were meant for the pupils from rural schools recommended as gifted.

Another law (1921) referred to the training in military high schools for officers. In general, we could say that the six of them (Iasi, Cernauti, Chisinau, Tirgu Mures, Predeal and Craiova) were exclusive. The entrance examination was difficult (about 30 candidates for a seat). The teachers were highly competent and thus better paid than the teachers in the other high schools. The pupils had enriched programs. These special high schools were also involved in various competitions in arts, sports, and academic abilities.

The law (Petre Andrei) passed on 27 May 1939 was the first one to make explicit reference to the education of gifted. The Ministry of National Education, and any other educationally oriented institution, could found special schools for "children endowed with and exceptional intelligence". The teaching staff of these schools was made up of graduates who had attended special post-graduate courses for such jobs. Their wages were 50% higher than their equivalents for the ordinary schools. The in-service training included special courses and research trips abroad. The pupils identified as gifted "on scientific bases" were to be trained according to "special methods and programs".

Romanian research on gifted ability began with an investigation of G. Aslan in 1915. He concluded that a differentiated approach was needed to deal with pupils according to their aptitudes and that special courses for notably gifted students should be established. Later, additional ideas and initiatives emerged from the field of giftedness research:

- survey of the situation in special education of gifted children abroad and in Romania, plans and suggestions for individualizing education based on the pupils' IQ (T. Dumitriu, 1928);
- the method of group selection for 11 year old gifted children, (L. Rusu, 1929);
- the responsibility of society to provide gifted students with a proper educational environment, (L. Rusu, 1929);
- performance selection and vocational guidance, (G. Stefanescu, 1929);
- collective set of tests, set up and patterned on the basis of scientific research, with a view to establishing a national pattern of IQ testing, (A. Rosca, 1930);
- setting up patterns for measuring and evaluating special aptitudes, for the first time in the history of psychology, (C. Parlog, 1932);
- taking as a starting point other countries' experience in the field, the author examines the need to group pupils according to their IQ and homogeneously displaying them in school grades, (I.C. Petrescu, 1939).

The war (1941-1944) brought about confusion and extreme right-wing political attitudes. In Romania, as in most European countries, education came to a stop. And than bad luck changed peace into the death of liberty, spiritual or otherwise. We are very much in the dark as to the full significance of the four decades (1945-1989) of socialist education but it is obvious that the gifted were put under a disadvantage from the start and stayed that way. The principle of the policy of education was "equality is identity", and in virtue of this guideline there was, on the one hand, a tendency to systematically level individuals and, on the other hand an insidious strategy of promoting the sham values who were finally imposed as leaders. The laws of education were extracted from the documents of communist party congresses. There was no local or national association for the education of the gifted; neither was there any involvement in international societies.

There was however a formal framework for the education of the talented pupils and the pupils with academic abilities. Mention should be made of music primary schools, art high schools, physical education and sports high schools, and special high schools for mathematics-physics, chemistry-biology, computer sciences, and history-philosophy.

There was also a network of academic and artistic competitions, going all the way from the gymnasium to the university, culminating in international Olympic competitions. The very highly gifted received intensive training in special courses as preparation for such competitions. Primarily in the fields of technical and artistic creativity, there was a marked tendency towards huge and propaganda-oriented actions, and were thus unprofessional in character. But, on a local scale, experts made up creativity-oriented teams, such as that associated with the Polytechnical Institute in Iasi.

The educational policy regarding people with high mental ability has not been supported by a scientific concern. The awarded pupils and their teachers were to be anonymous. The Ministry of Education did not subsidize any scientific research on giftedness. Research has made progress only in the fields of psychology and pedagogy of creativity. There has been a monograph study, brought up-to-date several times, about famous Romanian scientific and artistic personalities, about their birth-place and where they attended the compulsory school, covering the 18th and the 20th centuries.

Since the fall of the communist regime in 1989 the process of reforming the educational system has been under way. But the draft of the law of education, (May 1991), now undergoing public debate and analysis, has no special reference to any program for the education of gifted people. Art. 18, however, includes the principle of acceleration, to the effect that pupils and students with special professional competence are allowed to attend two or more years during one academic year. A large network of schools differentiated on categories of abilities (aptitudes) - not on levels, as yet - has come into being. Public debate addresses some other projects for the law of education, promoted by non-governmental groups. Nevertheless, no project has been passed as yet.

Despite this provisional state of affairs we have some clues to the future progress of reforms in Romanian education. As far as the education as such is concerned, steps have already been taken vis-à-vis: University autonomy, closer ties with research, cooperation with foreign educational institutes, and integration into the European educational system. Such tendencies are permissive for the education and research of the gifted. Several initiatives are already underway:

— the foundation in Iasi of the first local association for the gifted,
— research in the field of identifying gifted individuals (I.S.E.-Bucharest)
— research in the field of the curriculum for gifted,
— research in PhD programs,
— organisation of homogeneous training forms and groups,
— national conferences,
— contacts with international societies and organizations.

All these actions provide a favorable basis for a good educational policy and also for systematic research which could rightly be labeled the "accelerated and careful recuperation of the time lost".

References

Law 1904, *Rapport adressé à Sa Majesté le Roi de Roumanie sur l'activité du Ministère de l'Instruction Publique et des Cultes* par Spiru Haret, Ministre.

Law 1939, Law regarding the organisation and functioning of primary and normal education. Issued with the *High Royal Decree, Nr.2184/26 May,* 1939, chap. VI

Law 1971, Circulara Nr. 8345/1971, in *Buletinul Oficial M.I. Seria A, Nr. 3*/Nov. 1971

Law 1991, Legea invatamintului, Proiect, 1991, in: *Tribuna invatamintului, 12*/1991

Aslan, G., Improving personality through education, In: *Revista generala a invatamintului, Buc., Nr 6*/1915

Parlog, C. (1932) *The psychology of drawing; Testing and diagnosing talent in drawing*. Cluj., edit. Institutul de psihologie a Univ. Cluj.

Petrescu (1939), I.C. Methods used in studing individuality, In: *Cultura Romaneasca*. Bucharest.

Rosca, A. (1930). Testing the I.Q. and mental retardedness, In: *Cartea Romaneasca*. Cluj.

Rusu, L. (1929). The selection of gifted children, In: tip. *Cartea Romaneasca*. Cluj.

Stefanescu, G. (1929). Selection of performances and vocational guidance, In: *Cartea Romaneasca*. Cluj.

Theodosiu, D. (1928). Aptititudes and the selection of pupils for further or vocational studies, In: edit. *Librariei Socec*. Bucharest.

Tiberiu, B. (1981). Children capable of uncommon performances, In: *E.D.P.*. Bucharest.

The research on talent in Slovakia

Vladimír Dočkal, Jolana Laznibatová and Tomáš Kováč
Czechoslovakia

During the time of the "socialist build-up" in Czechoslovakia it had not been possible to examine the topic of special talents in children, as school policy was focused on "averaging out" the population. Only in the 60s – perhaps due to the Soviet-American fight for primacy in space – was the need for care of the gifted admitted. At this time the first classes for gifted pupils were established, and worked on the principle of providing a broader education in a particular subject, e.g., mathematics, foreign languages, and sport classes. Children were only able to develop artistic talents outside of the regular education system in the network of so-called basic artistic schools. Alternative forms of talent identification and development included various sports and artistic competitions, subject olympiads, and correspondence seminaries. All these forms of care were based on pedagogical empiricism.

The first sporadic research on child talent was begun in the 70s in the areas of sports (Bohemia) and intellectual abilities (Slovakia). The concentration of attention on the problems of the talented has been made primarily in Slovakia where, at the start of the 80s, a team of professionals was formed at the Institute of Child Psychology and Psychopathology. The extensive project of this team was based on the observation of a wide variety of intellectual (J. Laznibatová), physical (V. Palkovič), and artistic (V. Dočkal) talents, including the search for their common characteristics. This research provided important information about the idiosyncrasies of particular types of talent, as well as providing a better understanding of their similarities, both of which are necessary in the identification and development of a talent.

The research has focused on three facets of the problem of studying talented children:
1. theoretical analysis of the issues;

2. the problem of identification;
3. the problem of talent development.

Recently we have focused on some rarely examined types of intellectual talent (T. Kováč, E. Farkašová) and, generally, most attention is given to the interaction between a talent and creativity (T. Kováč, V. Dočkal). The following is a summary of some of the important findings and conclusions:

1. A gift/talent (both terms are used as synonyms) is formed on the basis of genetic dispositions. The significant factors affecting its emergence are the existing environmental conditions and the actual activity of an individual.
2. A gift regulates the activity of man. A gift is a characterization of personality and, in that sense, every human possesses some sort of gift. The gift/talent has a continuous character with respect to the degree to which it is expressed/developed. It may be quantified in terms of this criterion.
3. The qualitative aspect of a gift is based on its structure which conditions the so-called sorts of a talent (it is differentiated in accordance with the classification of human activity).
4. A gift is a multi-dimensional phenomenon. Its structure consists of a preconditioning constituent (somatic characteristics, abilities, skills, and acquirement) and an activating constituent (non-specific motivation, will attributes, personality orientation).
5. A significant part of both constituents of a talent is creativity (creative abilities, creative motivation). Insofar as creativity plays a significant and dominant role in the structure of a talent, we may speak about creative talent.
6. The particular components of either constituent of a talent possess a hierarchical structure, which is also related to the sort of talent in question. The most general components of a talent (which are involved in every kind of activity) are intellectual abilities, creative abilities, as well as performance motivation (a need for self-assertion). On the lower levels of this hierarchy there are the attributes which identify the specific talents. Nevertheless, some attributes may play a general role in relation to one activity (some talent with a special relation to some other activity).
7. The differentiation of talents begins in the course of development when the locus of a talent is gradually moved from its general characteristics to the special ones. Some common traits, however, are preserved. There are

certain sensitive periods for the development of particular types of a talent, but the actual course of development of a talent is strictly individual.

8. A significant part of any talent is the wide spectrum of child's interests and activities. In most cases, precocious specialization inhibits the development of a talent.
9. In the identification of talent it is necessary to examine all levels of the hierarchy that participate in a given activity. For example, if a young sport talent is to progress, s/he has to be endowed with at least an average level of intellectual abilities, and not only with the attributes that are specific to a given sport.
10. With talent development it is necessary to concentrate on several levels of a personality trait hierarchy. The good musician may be just the one who, in addition to good instrument technique, is able to develop, e.g. certain general artistic abilities.

These points serve as guidelines in developing the identification tests which can be used in individual counseling as well as for selecting the most eligible applicants for some specialized type of education. Tests in use are adaptations of foreign techniques as well as special purpose techniques developed by the team of researchers.

In the development of talent we focus on a well-balanced stimulation of both the preconditioning and the activating constituents of a talent, mostly in programs for the development of creativity. The significant aspect of these programs should be anxiety reduction, because anxiety inhibits the free development of a talent. The research in this area is forthcoming (M. Matejík). We are also examining the Czechoslovakian school system, mostly on the effects of the above mentioned selective classes. It seems that the didactically positive aspects of these special classes are coupled with several negative aspects, at least as far as personality development is concerned. Thus, the pupils are unambiguously more individualistic, but are wrapped up in themselves, they lose motivation for work more rapidly than pupils from heterogeneous classes, they are more liable to neurosis, family separation (e.g., at sports boarding schools) creates many problems, and so forth. All these issues are topics for further research by the workers from the Research Institute of Child Psychology and Psychopathology in Bratislava.

Programmes for highly able children in Austria

Barbara Grillmayr
Austria

A survey of different models of gifted education in Austria will be given. The following topics will be dealt with:

– the role of gifted education in the Austrian school system
– political influence on the development of gifted education
– different kinds of support for gifted children
– future perspectives

THE ROLE OF GIFTED EDUCATION IN THE AUSTRIAN SCHOOL SYSTEM

In Austria all children attend the same type of primary school from the age of six to ten. At the age of ten pupils must decide either for Grammar School ("Gymnasium" for intellectually able students) or Secondary Modern School ("Hauptschule" for those who are not conspicuously intellectually able). A second decision must be made at the age of fourteen, either:

– to learn an occupation and finish training at age 17 or 18,
– to remain at Gymnasium until age 18 and possibly go on to university, or
– to attend a top-level vocational school until the age of 19, and still to have the possibility of going on to university.

The Grammar School originally gave sufficient support to highly able pupils. However, the percentage of pupils attending Gymnasium has been constantly increasing, and in certain districts of Vienna up to 80% of children attend Grammar Schools. Thus, possibilities for supporting especially gifted pupils in Grammar School have been drastically reduced. That is why one of the main educational aims, namely support for the disadvantaged and for the gifted, could no longer be achieved. Therefore special programmes for the advancement of gifted students had to be established. Most of the programmes for highly able children that have been started in Austria are offered in addition to normal education, i.e., enrichment. A second alternative, acceleration, is limited by law. Grade skipping is one example of the problems in the field of gifted education. To this day skipping is only possible for students who have been abroad for a year, and who wish to rejoin their former class upon returning to Austria. A third alternative, grouping (measures taken to create homogenous groups), is only rarely practised. For a detailed explanation of the terms "enrichment", "acceleration" and "grouping" refer to Geuss & Urban (1982, pp. 100-103).

POLITICAL INFLUENCE ON THE DEVELOPMENT OF GIFTED EDUCATION

During a Socialist majority rule under Bruno Kreisky between 1970 and 1984, extensive programmes to help disadvantaged learners were conducted. Entrance requirements for Grammar Schools were removed, which enabled a large percentage of children to attend the "schools for the gifted". Apart from a few competitions, the promotion of gifted students was neglected.

In 1986 the Speaker for Education for the Conservative Party, Gerhard Schäffer, declared the advancement of gifted children to be a most urgent matter. Following this declaration, gifted education became a major issue in Austrian educational policy. As a consequence the following steps were taken.

The president of the school district of Salzburg, Gerhard Schäffer, started a series of school experiments. At the same time other states, especially those with conservative majorities, developed their own advancement programmes. In order to support these efforts, an "Association for the Promotion of Gifted Students" was set up.

Several of those school experiments turned out to be a great success. As a result, a law concerning gifted education was enacted on June 9th, 1988. Thus advancement programmes for gifted children were included into the Austrian school system by law.

DIFFERENT KINDS OF SUPPORT FOR GIFTED CHILDREN

According to this law Grammar Schools have to offer "additional courses for exceptionally interested and gifted students" from 14 to 19 years old. There is no curriculum laid down, and course leaders are free to choose any topics. All other types of schools are entitled to start the same courses. Students from any Grammar School in a district between the age of 14 and 19 years are allowed to take part in every course offered. This means that pupils between 14 and 19 years of age might attend the same "lessons" and that students in one course might come from several different schools. An opening number of 10 students per course and an even lower one for practically oriented classes with laboratory work guarantees optimal working conditions for especially interested and able pupils. The groups meet for two hours every week. In contrast to the earlier experimental programmes there are no special entrance requirements. The fact that the pupils volunteer for the courses is regarded as a guarantee of their interest and their ability. Rapid pace plus complicated content would be expected to automatically exclude students with insufficient interest or ability. The topics of these courses range from archaeology, electronic music, English media, and media politics to immune biology, computer graphics, data-processing, and the special theory of relativity. In the 1989/90 school year the first courses within the scope of this law were offerred in the school district of Salzburg. Some of the results of an "evaluation" of the programme were the following.

Statistics of the School Board and questionnaires filled in by the students at the beginning and at the end of the first year provided the following details about the participants and their attitudes:

- average age: 16.3 years;
- average mark in the last report: 1.89 (where 1 is the highest mark possible);
- average mark in the subject most closely related to the course attended: 1.87

A comparison of the marks (the general ones and the marks in the subjects most closely related to the attended courses) leads to the conclusion that the courses do not really promote the particular talents of students. The courses support students' general interests but not their particular high ability in a special field. Thus, the courses have much in common with the Saturday Clubs in England and with the courses offered by the "Deutsche Gesellschaft für das hochbegabte Kind" (Cf. Heinbokel, 1981).

When students were asked about their reasons for taking a course the following answers were most often given.

1) improvement of their knowledge of the particular subject: 57% (66%)
2) improvement of their general knowledge: 42% (80%)
3) gaining some foreknowledge for their prospective career: 32% (21%)
4) the challenge of something entirely new: 32% (38%)
5) the chance of working independently: 55% (41%)

The first number gives the percentage of students who gave that answer for why a course was chosen. The number in parentheses is for the percentage of students who reported that the specific goal was in fact fulfilled.

These numbers suggest that the students were generally quite happy and satisfied with the courses. The fact that their expectations regarding the utility of the courses for their future career were not fully met is not surprising, since the courses are not intended to provide advance knowledge for future jobs or professions. In any case, most of the participants did not expect such information or training. What does have to be improved is the chance for students to work independently.

The students and the teachers of each course also wrote short reports about what they had worked on and about their experience with this new type of course. These reports indicate that students really appreciate the chance to participate in this programme. However, some of the students and teachers found it difficult to work together with others whose background knowledge differed greatly from their own. This problem results from the general age limits (14 -18) for each course, from the absence of entrance requirements, and from the fact that "exceptionally interested" does not necessarily mean "exceptionally gifted", nor does "exceptionally gifted" necessarily mean "exceptionally conscientious". Some course leaders complained that interest flagged at the times when the normal school subjects were being tested.

Apart from these enrichment courses for the exceptionally gifted there are some other programmes, which can be briefly described.

- Advancement of musically talented children (age 6-10): An enrichment programme which aims at finding and promoting musically gifted pupils. It is offered at Salzburg primary schools. The "Mozarteum" (Salzburg university for music) and the "Salzburger Musikschulwerk" have helped in developing the programme and both institutions now assist pupils and teachers involved in it. Children, parents and teachers decide whether a child shall take part in the programme. The programme's target is certainly not the creation of professional musicians, which is sometimes the case with projects fostering musical talent at an early age (cf. Feger, 1988, p.88).
- Promotion in computer studies for students aged 10 to 14 and living in rural areas.
- Special training for athletes: students aged 14-19 attend a special school which offers the same curriculum as a normal Gymnasium but also devotes a great deal of time to sports tuition and training. As the students need a great deal of training time they have one additional year at school. They have 25 lessons of academic work per week, and 20 lessons of specialist training.
- Olympiad competitions in mathematics, physics, and chemistry.

In two Federal States additional language tuition is offered in Latin, Greek, English, French, Russian and Spanish and a competition is held in March.

FUTURE PERSPECTIVES

As the present courses for exceptionally gifted and interested pupils at Grammar Schools have met the approval of teachers, students and parents they will be continued and developed although there will have to be some changes in entrance procedures. A further step will be the start of similar classes at other types of schools. All teachers are and will be encouraged to support gifted children in normal classes as well. Teachers should be able to recognise and foster special abilities of their pupils. They should create an optimal learning environment suitable for pupils of different abilities because every regular class is heterogenous regarding abilities. As George (1990) noted, "a teacher needs to be flexible, to increase diversity of teaching methods, learning experiences and needs to respect, encourage and support the able child." That is why further seminars on this topic will be offered. The culmination of discussion in Austria to date has been the regional conference of the World Council in Salzburg. A report of this conference (entitled "Needed – the Gifted!" and published bilingually in English and German) is available from the author of this article.

References

Feger, B. (1988). *Hochbegabung. Chancen und Probleme*. Bern: Hans Huber.

George, D. (1990). The challenge of the able child, *Cambridge Journal of Education, 20*, 175-182.

Geuss, H. & Urban, K.K. (1982). Hochbegabte Kinder. In Wieczerkowski W. & H. zur Oeveste (Eds.). *Lehrbuch der Entwicklungspsychologie,* pp. 85-110 Vol.3. Düsseldorf: Schwann.

Heinbokel, A. (1981). Modelle der Hochbegabtenförderung in den USA, in Israel und in Grossbritannien. In W. Wieczerkowski & H. Wagner (Eds.). *Das hochbegabte Kind.* pp. 189-200. Düsseldorf: Schwann.

Different pupils - different teaching

Mogens Jansen
Denmark

There is always a need to consider possibilities for making teaching more efficient. Both the individual and the society must be able to develop through optimal utilization of the resources available and in the best possible harmony. Naturally, the objective is to attain the best possible personal and social development. The means to achieve this should be discussed and should be constantly evaluated in the light of new social possibilities.

In a world which is on one hand being internationalized and where on the other hand the traditions of individual countries, areas, regions become still more distinct, it is important that education is not only internationally oriented but is also based on and maintains bonds with the cultural environment within which children grow up and live.
It is necessary:

- to start from the different systems of education;
- to maintain the respect for the starting point and the possibilities of the individual;
- to maintain and expand democratic ideals; and, to the greatest possible extent;
- to work towards a mutual inspiration within a community characterized by differences (Skov, 1984; 1986).

TRENDS IN PERFORMANCE IN VARIOUS SCHOOL SUBJECTS

It is not possible to describe the status of all subjects on all levels in school in exact figures. There simply are no tests to do this (Lau & Pagaard, 1990). In some subjects very precise research has been carried out; in other subjects a teacher evaluation after the 9th school year provide fairly reliable descriptions.

The only subject that has been thoroughly looked into at most grade levels is reading. Since the 1960s repeated measurements, both through very small intensive studies and through very large nation-wide surveys, have indicated some general tendencies in Denmark (Dalby et al., 1983; Jansen & Kreiner, 1986). These can be summarized as follows:

- Children are learning to read later than a generation ago, but within a few years in school they become essentially better readers than was previously the case.
- The number of books being read is increasing.
- The diversity of the acquired texts is increasing (Dalby et al., 1989).
- The decrease in the number of pupils, which has been evident for a number of years, does not fully correspond to the growth in the "book consumption" that can be recorded.
- Changes in the distribution in reading competence measured among all Danish children can be described as a "Matthew tendency " (cf. Hansen et al., 1987), where children's performance scores which are at first normally distributed become "squeezed" towards the extremes. Over time, this results in an increasingly tri-modal distribution, with a small peak of very poor as well as of very good readers. In the 1970s the squeeze effect meant that there were more good readers than in the 1960s. The most recently observed tendencies at the beginning of the 1980s indicate a new squeeze effect, resulting in a further polarization between good and weak readers. The tendency continues and with advancing pupil age the group of good readers in particular is growing. On the other hand, it looks as if a numerically modest, yet still too large group (in some cases 15%, in other cases more) risk getting "tied off", isolated from the standard educational processes.

During this period a massive integration of earlier separated pupils has taken place. These pupils now follow the ordinary education in much larger numbers than before. However, this group of pupils is not as large as the growing number of pupils whose performance is below normal (Dalby et al., 1989).

Approximately the same tendency found in reading has been found in mathematics (Foss Hansen, 1983). The findings of this research are supported by the somewhat less systematic annual teacher evaluations which take place at the end of the 9th school year.

In spelling there seems to be a relative increase in the number of the highly performing pupils (Alerup et al., 1991). This study, which compared spelling competence in 1979 with 1990 is continuing; further results are expected to be published in 1992.

An assessment of performance in written composition is restricted to the results of the teacher evaluations made at the end of the 9th school year. The tendency here seems to correspond to the tendency shown in spelling.

At the same time there is a declining birthrate in Denmark (Jansen & Kreiner, 1985). This means that the educational institutions and trades in need of very competent pupils are having difficulty in getting the same student enrollment as they used to. In particular, the middle level of education will have difficulties in obtaining sufficiently competent pupils. To fill the positions available, they are approaching the threshold of the group of pupils who have specific difficulties in, among other topics, reading. The trades and educational institutions recruiting pupils from the weakest areas, as far as literacy is concerned, will certainly also have very considerable problems. Yet, at the same time, it is realistic to expect increasing requirements in literacy from many educational institutions that used to be "practically oriented", e.g. the trades (Jansen, 1987).

Taken together, the Matthew tendency, a declining number of pupils, and increasing demands of society mean that it has become more difficult than ever to teach all pupils "in the same way".

STREAMING OR NON-STREAMING?

Wahl Andersen (1991) describes educational development in Denmark from the first Danish Act regarding public education in 1814 up to the present day. Without exception, the development has passed from a streamed education towards a non-streamed education. Since the 1970s the majority of pupils have remained in the same class group from grade 1 (the age of 6-7) up to and including grade 9 (the age of 14-15), yet in such a way that in some subjects it is possibe to choose courses on different levels. The number of pupils enrolled in a separate special education programme is constantly declining.

In Denmark the development of the comprehensive school has taken place in the context of ongoing debate. In its present form the non-streamed school is the reality for almost all children up to at least the age of 14. This tendency continues up to and including about age 16. Even though the debate continues, it seems unquestionable that streaming after the age of 14-16 will be maintained. In contrast to only 5 to 6 years ago, very few voices urge a comprehensive school beyond the age of 15. There are correspondingly few advocates of a return to streamed schooling. The non-streamed school could be said to finally have the upper hand from the start of schooling to the end of the 9-year school.

ALL PUPILS MUST BE CONSIDERED

A few generations ago the educational system tried to educate an élite for the official class and a somewhat larger group for the "non-literary" trades, and leaving a large group without any special education (Dalby et al., 1983). The social development in a country with a structure like that of a Nordic country has prevented the continuation of this tendency.

A generation ago it was even more strongly argued in favour of considering especially the weakly performing groups of pupils. As a caricature it was said that "equality" was rated higher than "competence".

Today there is practically total agreement that the objective of education is that all groups of pupils, and each individual pupil, should be able to develop optimally. This also applies to the specifically gifted children. The earlier way of thinking claimed that this group of pupils "would always do well", a theory which has proven untenable (Rabol Hansen et al., 1992).

It is emphasized in Denmark that a differentiated education is not an individualized education. A great deal of importance is attached to the fact that education takes place in a social context and that the goal is the versatile development of each individual. The individual enriches and inspires the community while the community supports the individual. This is considered to be a primary principle in the Danish school system.

But the problem remains: How do we individualize instruction? There is general agreement that the problems cannot be solved by individualizing the group of pupils. There seems to be general agreement that instructional individualization is a valuable concept, but how do we put it into practice? (Jansen & Ziegler, 1990; Ziegler, 1991; Lundgreen, 1981). Among other things, these works list the following three factors: Teaching materials, the organization of the education, and educational activities.

Two important research results are the following:

1. The pupils' working method and the teachers' understanding of this way of working. This seems to be decisive in order for the individual pupil to acquire a work style appropriate to the emerging social changes, e.g., the breakup of countries and cultures, and not least due to a constantly changing technology.
2. The personality of the teacher appears to be a very significant factor. Some teachers are capable of motivating and engaging the individual pupil and the class in school activities. But how to describe the "personality of the teacher" in detail is another matter. This is a research problem of enormous dimensions.

RESEARCH AND DEVELOPMENT

For several years very comprehensive research and development programmes have been carried out. They are followed very closely by the research institutes - in the hope of discovering the origin of the problem and to find a way to (more or less) set it right (Frey et al., 1991; Jansen & Ziegler, 1990; Ziegler, 1991).

Within these fields a number of research projects are now in progress in Denmark - far more than has ever been the case before. Some of these projects are subjected to a systematic evaluation; most of the projects will be finished during the period 1991-92. A few projects will be continued for a year or two. Hopefully, by that time it will be possible to consider all the pupils more thoroughly than now - for both human and social reasons.

The tendency clearly indicates that not only the disabled children, not only the children in less privileged circumstances, the children at risk, but all children must be offered the opportunity to develop optimally. This is the perspective in which the education of the exceptionally gifted children will be seen - at least this is the direction in which the tendencies point.

References

Andersen, W.M. (1991). *Differentieringsformer, belyst gennem udviklingen i folkeskolen* (Forms of Differentiation, lucidated through the Development of the Comprehensive School). Uddannelse, 24(7), 419-431.

Allerup, P.; Jansen, M.; Johansen, J.; Klewe, L. & Kreiner, S. (1991). *Stavefærdighed - nu og for 11 år siden* (Spelling Competence - now and 11 Years ago). Copenhagen: Danmarks Pædagogiske Institut.

Dalby, M.A.; Elbro, C.; Jansen, M.; Krogh, T. & Ploug-Christensen, W. (1983). *Bogen om læsning I. - forudsætninger og status* (The Book of Reading I - Background, Conditions, and Status). Copenhagen: Munksgaard/Danmarks Pædagogiske Institut.

Dalby, M.A.; Elbro, C.; Jansen, M. & Krogh, T. (1989). *Bogen om læsning II - set i et udviklingsperspektiv* (The Book of Reading II - seen in a Perspective of Development). Copenhagen: Munksgaard/Danmarks Pædagogiske Institut.

Frey, B.; Lau, J.; & Skov, P. (1991). *Specialundervisningens mange ansigter* (The many faces of special education). Copenhagen: Danmarks Pædagφigiske Institut.

Hansen, K.F. (1983). *Færdigheder i regning/matematik - diagnostiske prφver og lærebogssystemer* (Skills in Arithmetic/Mathematics - Diagnostic Tests and Textbook Systems). Copenhagen: Pædagogiske forskningsrapporter: Den gule serie.

Hansen, M.; Kreiner, S. & Pagaard, P.E. (1987). *Dansk - et fag i krise?* (Danish - a Subject Undergoing a Crisis?). Horsens: ålφkke.

Hansen, R.; Jansen, V.M.; Lau, J. & Robenhagen, O. (to be published in 1992). *Tidlige læsere* (Early Readers). Copenhagen: Danmarks Pædagogiske Institut.

Jansen, M. (1987a). *The fight against illiteracy — described from a Danish starting point.* Rapport udarbejdet til Undervisningsministeriets internationale kontor (Report prepared for the Danish international section of the Ministry of Education and Research). Copenhagen: Danmarks Pædagogiske Institut.

Jansen, M. (1987b). *A Little About Language, Words and Concepts - or What May Happen When Children Learn to Read.* Copenhagen: Landsforeningen af Læsepædagoger.

Jansen, M. (1991). *Changing Reading Demands in a Changing World.* Budapest.

Jansen, M. (1991a). *Improved Reading Instruction and More Problems: The Phenomenon Is International. But What Do We Do About It?* Las Vegas: International Reading Association (36th Annual Convention).

Jansen, M. (ed.) (1991b). *Undervisningsdifferentiering — ja, men hvorfor og hvordan?* (Educational differentiation — yes, but why and how?). Copenhagen: Danmarks Pædagogiske Institut.

Jansen, M. & Kreiner, S. (1985). Når der bliver færre bφrn i skolen — om prognoser, hφjteknologi og pædagogiske konsekvenser (When the Number of Children in School Decreases - About Prognoses, High Technology, and Educational Consequences). Copenhagen: Danmarks Pædagogiske Institut.

Jansen, M. & Kreiner, S. (1986). *Hvor meget dårligere eller bedre læser bφrn i dag? - en opgφrelse over læseniveauet i 2.-5. klasser for årene 1976-84, sammenstillet med tidligere år* (How much worse or better is the Reading of the Children Today? - An Account of the Children Today? - An Account of the Reading Level of Grade 2-5 Pupils for the Years 1976-84, as compared with Previous Years). Copenhagen: Pædagogiske forskningsrapporter nr. 36 (Den gule serie).

Jansen, M. & Ziegler, M. (Eds.) (1991). *Tid til dansk 1* (Time for Danish 1). Summarized in *Reading Development — A Question of Quantity & Quality*. Copenhagen: Danmarks Pædagogiske Institut.

Lau, J. & Pagaard, P.E. (1990). *Gruppeprφver i læsning — en oversigt* (Group Tests in Reading - A Survey). Copenhagen: Danmarks Pædagogiske Institut.

Lundgreen, U.P. (1981). *Model Analysis of Pedagogical Processes*. Lund: Liber Läromedel.

Skov, P. (1984). *Delt eller udelt skole?* (Streamed or Non-streamed School?). Copenhagen: Danmarks Pædagogiske Institut/Folkeskolens forsφgsråd.

Skov, P. (1986). Mixed Ability Teaching. *Scandinavian Journal of Educational Research, 30,* 85-94.

Ziegler, M. (Ed.) (1991). *Tid til dansk 2* (Time for Danish 2). Copenhagen: Danmarks Pædagogiske Institut.

Research on and education of the gifted youth in Yugoslavia

Slavica B. Maksiç

Yugoslavia

INTRODUCTION

There is a long tradition of interest in giftedness and gifted individuals in Yugoslavia. As early as 1910 Paja Radosavljevic devoted a chapter to the nature and development of giftedness in his book "Introduction to Experimental Pedagogy". While Terman was beginning his study of giftedness, Yugoslav journals published reports on the treatment of the gifted in the U.S.A. and in Germany as well as reports on methods for selecting gifted students. However, it is only in the seventies that research on giftedness was systematically begun. In this period there was also a significant change in community attitudes towards the gifted. There were increasing calls for special action to provide favourable conditions for the development of gifted individuals and for their subsequent participation in society.

Increase in and differentiation of social interest in gifted children and youth can be observed on several levels; from research projects and published reports and symposia on the gifted, to educational programmes provided for gifted pupils and social conditions in which the gifted can expect support.

FROM 1960 TO 1969

At the start of the sixties the first articles on creative thinking and performance began to appear. Research on creativity was from the beginning associated with school performance, e.g., in engineering, art, and language. By the end of the decade the research methodology was established and possibilities for determining and measuring creative abilities were being discussed.

During the sixties schools began to accept that children differ in ability and that a differentiated approach should be taken for different categories of pupils. Questions regarding how to motivate the advanced pupil and attitudes towards these pupils began to be addressed. School policies also began to change. Special awards were introduced for those doing extraordinarily well in school, both overall and in specific subjects. A law passed in 1959 required schools to organize corrective instruction to help backward pupils and to provide extra classes for advanced pupils. In the same period a system of contests in various subjects (math, natural sciences, languages) was set up, progressing from the local school to national levels, and provided for advancement to international competitions.

In addition to regular schools, institutions with an educational character began to appear, intended for pupils interested in technology. Since the 50's there have been clubs for young engineers and a National Technology Club intended to arouse interest in engineering and technology among elementary and secondary school pupils. Hundreds of thousands of students as well as teachers participate in these clubs and in contests organized from the local to national level in 25 fields of study.

The scientific periodicals in this period showed an interest in topics such as the status of the gifted and talented in school, and published translated foreign articles on the topic of giftedness. Articles stressed the need to identify and provide assistance for gifted students. As a result of this growing interest, some practical steps began to be taken.

FROM 1970 TO 1979

The number of publications doubled in the seventies. Work on creativity continued, examining it's nature, measurement techniques, and ways of improving creativity in students. An increased interest in giftedness is evident, and most work deals with the identification of gifted individuals, their schooling, and vocational guidance. In the early seventies several employment bureaus began studies of the characteristics of gifted youth. This research was carried out by vocational guidance services with the aim of providing the gifted with opportunities to choose an adequate vocation and corresponding education.

In 1969 the Republic Fund for Crediting Students and Pupils in Croatia began granting certificates of credit to students. These were based on recommendations from teachers, on psychological tests, and on school proficiency criteria. These certificates could be used when applying for a scholarship, entering higher levels of education, or for entering specialized schools for the talented.

Vocational guidance for the gifted proved to involve many practical problems, and this served to initiate research on the topic. Issues regarding the identification of the gifted became a subject not only for employment bureaus but also for scientific institutions. In 1971 the first symposium on giftedness was held, and dealt with the vocational guidance and schooling of gifted students. This was followed by a symposium on education of the gifted in secondary school, and one on the identification of the gifted and how to provide adequate developmental support.

Additional regulations were introduced in the seventies concerning the education of gifted students in elementary and secondary schools. Schools were required to offer organized extra work, the formation of various "scientific groups", optional lectures, and accelerated education programs to students who showed higher than average ability or interest in school subjects. This led to the decision to have gifted pupils attend classes in ordinary schools but to offer them additional courses and activities in addition to the regular classes and school curriculum.

The value and justification for some of the forms of education for the gifted were frequent topics of research. Topics included the effects and success of acceleration, of differentiation and individualized teaching, extra-curricular activities, the social adaptation of the gifted, and characteristics of the teachers of the gifted.

This period is also marked by the establishment of several extra-school educational institutions. For example, the aims of a mathematics club are to teach students, popularize mathematics and science generally, organize contests, and cooperate with teachers of mathematics in regular schools. Other specialized schools were established. In 1973 a school of music was founded for those who displayed an extraordinary gift for music. In addition to the regular teaching staff, university professors are often engaged in working with the students.

FROM 1980 TO 1990

The number of publications and research projects again increases in this decade. Creativity remains an important topic for study, but it is evident that creativity is increasingly associated with education. Creative and analytical learning, creative style in teaching, pupils' creativity in school, and the impact of teaching on creativity are among the topics examined. As in the previous decade, most of the work deals with the gifted, and mostly with their characteristics, identification, and schooling. There is a shift away from vocational guidance to professional development.

There is an increase in theoretical analyses of the notion of giftedness, an increase in research on the characteristics of the gifted, and new identification procedures are devised. In 1988 the Institute for Psychology in Belgrade began a project to examine the methods and instruments for identifying giftedness. At the same time the psychological analysis of the gifted was begun with the aim of providing a theoretical and methodological framework for understanding the developmental characteristics of high intelligence.

The study of giftedness extended to topics beyond the school environment. The conditions which society provides for the gifted and the role of the gifted in society are examined, as well as other essential features such as the family environment. One important project is currently examining the effects of public identification and treatment of the gifted, to determine the social consequences of labelling on the gifted and on the social support they obtain. The education of the gifted still receives the most attention. Research on extra-curricular activities, acceleration programs, and differentiated and individualized teaching continues. Other research examines the value and effect of contests, awards, clubs, and science centers. A new topic which is beginning to receive attention is gifted underachievement.

The eighties are also marked by the establishment of the Organization for Scientific Research in Zagreb (1982), the Belgrade Research Center (1988) and the Petnica Science Center (1988). The latter is an institution open throughout the year, and allows students and teachers to attend a variety of seminars and courses as part of their extra-curricular and free time acitivities. In addition, many institutions have been founded to provide various types of support (primarily financial) to gifted children in the course of their schooling and advanced training. Some organizations and communities form special funds to grant scholarships for the gifted in order to direct them towards scientific research work. These efforts are supported by the higher scientific institutions. Under the patronage of the Servian Academy of Science and Art a foundation was established in 1987 to promote the development of scientifically and artistically inclined youth. It's primary activities include granting scholarships, financing study trips in Yugoslavia and abroad, and publishing the work of it's members.

SINCE 1990

By the end of the eighties a variety of programs for the gifted had been set up. These aim at establishing and improving methods of working with the gifted. Educational policy continues to promote specialized education for the gifted. One recent innovation involves establishing specialized secondary schools and classes for educating the highly capable students. Admission to these specialized schools is determined solely by proficiency tests and success in school. This program is still in the experimental stage. Such status is also given to other programs being carried out throughout the country. The network of institutions, in addition to schools, intended to develop scientific interest in youth, and to work with the gifted, continues to spread. The fact that research is being planned which will, in a more complete manner and with a more complex methodology, attempt to study the value of instructional methods for the gifted is encouraging.

A new era in education: England and Wales in the 1990's

Johanna M. Raffan & Michael Short
United Kingdom

The last decade has seen enormous changes in the U.K. education system. In particular in England and Wales the introduction in 1981 of the Special Education (Warnock) Report defined legally the ways to address students' special needs. Unfortunately, this occurred for only one end of the intellectual spectrum. "We did not regard the problems of highly gifted children as falling within our remit, except insofar as these problems may result in emotional or behavioural disorders similar in effect to the problems of other children with whom we are concerned". The 1988 Education Act has resulted in even greater changes. The full impact of this Act is still to be experienced and understood, as parts of it have not yet been implemented. A large number of facets of the educational world are affected by the act: Local Managements of schools; Open Enrolment; School Governors; National Curriculum along with its associated testing and assessment procedures; Testing and Assessment; Grant Maintained Schools; City Technology Colleges. However even with so much of the education system enshrined in law the needs of gifted children have yet to be encompassed. Therefore organisations concerned with the able children in this country have had to adopt policies and strategies to meet the needs of able children within the existing legislation.

Local Management of Schools means that local authorities have to devolve to schools at least 85% of their expenditure on schools. The implications of this move are that schools are more autonomous. But, every child of statutory school age is worth a certain amount of money; therefore, the more pupils you have the more money the school gets. This policy is linked with Open Enrolment whereby schools no longer have to adhere to a catchment area. This gives parents much greater freedom of choice but also means that some schools could become very over-subscribed.

School Governors: Every school has a Governing Body made up of Local Authority Representatives of the local community, business and industry, and parents and teachers. This body's powers have increased significantly and they are now responsible for working with the Head teacher in:

— setting the general aims and ethos of the school,
— determining a school curriculum policy,
— conforming to national requirements,
— making decisions about what is taught,
— making decisions about delegated budget,
— determining priorities for spending on books and equipment,
— interviewing and appointing staff.

This increase in the Governors' responsibility and their greater involvement in the running of the school means that they must be convinced that gifted children have special needs if they are to ensure that the gifted are catered for in the ordinary classroom. To help Governors understand this need NACE has produced a booklet on "The Able Child and the School Governor". This deals with the major issues of identification, school policy, resources and training.

The National Curriculum consists of 10 subjects which all children must study at school. English, Mathematics, and Science are the core subjects. Design Technology, History, Geography, Music, Art, Physical Education, and a Modern Foreign language for 11-16 year olds, are the foundation subjects. In addition, there is Religious Education. All subjects have objectives or goals setting out what children should know and be able to do at each stage of their schooling. These objectives are called "attainment targets". For each subject there will also be a description of what children should be taught to help them achieve the attainment targets. These are called "programmes of study". There are four stages for different age groups known as key stages. These will help parents to know what their children should be learning at various stages.

Key Stage 1: 5 - 7 years
Key Stage 2: 7 - 11 years
Key Stage 3: 11 - 14 years

Key Stage 4: 14 - 16 years

The introduction of the National Curriculum has provided a national framework for education. It helps to meet the needs of all children, encourages all children to do the best they can, and enables parents to know more about their child's education and progress. Even though the National Curriculum ought to cater for the needs of all children, there may still be difficulty coping with children at the extremes of the abilities spectrum. Therefore additional help and support is essential for the very able child if he is to fulfil his potential.

Linked with the National Curriculum is a 10 point scale to cover the years 5-16. Children will progress through the 10 levels at different rates and their progress will be assessed at the end of each Key Stage. An average child will reach level 2 at 4 years, level 4 at 11 years, level 6 at 14 years, and level 7 at 16 years. The main aim is to help children progress at the rate which is right for them. This progress will be measured partly by teacher assessment and partly by standard assessment tests. Again, this presents problems for gifted children, many of whom will be at higher levels than their age group and this raises the question of acceleration. Given that the National Curriculum is central to our education system, it is essential that UK organisations concerned with the education of more able pupils focus their work upon the implications of this for the more able child, lest it be assumed that the National Curriculum in itself is a sufficient mechanism. Experts in the field in this country feel that most children should be catered for in the ordinary classroom with a differentiated curriculum. Their level of maturity and emotions are often not matched to an older age group although some of their academic abilities may be in advance of the National Curriculum requirements for their age. The course content of some of the foundation subjects is beyond the experiences of a younger child who is accelerated to a higher age group, despite the fact that the child may well be able to cope with the level of work. Although, we would agree that in exceptional circumstances acceleration may be necessary.

Thus as the act recognises that there are more able and highly gifted children in schools who will attain level 8,9 or 10 by or even before the age of 16, the issue regarding a differentiated curriculum must be addressed. NACE has for many years provided resources for Enrichment and Extension of the curriculum. Now it is addressing this deeper issue and has already published a booklet on "Differentiation and the Able Child" and is now looking at examples of good implementations.

The National Curriculum is mandatory for all children aged 5 -16 in state schools. This raises a problem concerning relationships between state education and private education. The NAGC survey, "According to their Needs", found that not all schools attempted to identify able pupils. This seemed especially so with independent schools where selection by ability was frequently regarded as sufficient (cf. one Head's definition of giftedness as "The ability to win a scholarship"!). Selective entry really assesses high achievers. Where there was no selective entry it was also more likely that specific efforts were made to identify and provide for the gifted. In general, the survey found that provision depends more upon the goodwill of individual teachers. In other words, it is something of a lottery. The Government puts 63 million pounds into the Assisted Places Scheme – whereby each year some 6000 pupils are able to receive independent school education. Supporters of the scheme believe that this is a major provision for able children. Critics note, however, that many of these children would have done just as well in state schools and they are equally concerned that the removal of the ablest pupils can lower the attainment level expected and gained by children within the state system.

Fundamental to all this is the teacher. Some local authorities have over the years tried to help and support teachers by providing advisers or advisory teachers to work in the field of Gifted Education. However, the NAGC survey carried out on behalf of the DES showed that only 5 out of 105 authorities designated a person whose sole responsiblity was for able children and one of them has since been made redundant through lack of funding.

Teachers therefore will need support from other quarters. CATE, the body responsible for accrediting teacher training courses, has as one of their criteria the following:

> "the capacity to identify gifted pupils and pupils with special needs or learning difficulties and to understand ways in which the potential of these pupils can be reached".

However a NACE sample survey of the 92 Colleges providing initial teacher training showed that out of 34 Colleges only 2.5% had courses specifically related to the needs of the able child.

Yet, the route to further progress in education for the able child *must* be that all initial teacher training courses should provide teachers with strategies and styles to meet the needs of these children, whilst those already teaching are given In-service Education. The latter has been made possible for the first time this year through the Government GEST budget (money provided by the Government to local authorities for In-Service Education) that allows schools to bid for money for "training to meet the needs of gifted children."

Measuring divergent abilities

Jasmina Šefer[1]
Yugoslavia[2]

INTRODUCTION

Creativity in general is very difficult to define precisely, and is dependent on an author's theoretical perspective. The definition usually refers to some of the manifestations of the phenomenon, such as the creative product, creative process, personality traits of the creative individual, or the external conditions favorable for the development of creativity. Defining creativity in children is especially difficult since the child is not mature and does not decide to produce a creative product. Rather, the child creates spontaneously, most often through play, which makes it difficult to distinguish the unintentional achievements from those which are a result of actual creative dispositions. Moreover, the nonlinear developmental changes in a child's personality makes it difficult to predict future creativity on the basis of creativity in childhood and youth. As a working definition, we can state that "children's creativity results from specific abilities and preferences, actualized in a favorable immediate environment through play-like activities, and may (but not necessarily) have as a consequence an interesting and unusual product".

This study examined one prerequisite of creative production. Specifically, the cognitive component of personality known as divergent thinking was examined,

---

[1] Jasmina Šefer, The Institute of Education Development, Beograd, Yugoslavia

[2] This paper is based on an MA thesis titled "Creativity in Young School Children - Their divergent Thinking Abilities and Personality Traits as evaluated by Tests and Other Methods". The thesis was successfully defended in 1987 at the University of Beograd Humanities School, before a committee composed of the following members: Prof. Ivan Ivić, - mentor, Prof. Vera Smiljanić, and assistant Prof. Panta Kovačević.

since many psychological studies show a specific and significant contribution of divergent thinking to creative production. Divergent thinking is often contrasted with convergent thinking. Divergent thinking tends to move beyond the limits of an assigned problem or the familiar frame of reference, and explores unfamiliar options in a relaxed but curious fashion. It engages the prelogical processes to a greater extent, characterized by greater associativity, imaginativeness and intuitiveness. Divergent thinking may produce an inadequate, bizarre answer, but convergent logical processes, especially those engaged in the preparation, elaboration, evaluation and verification phases, may serve to correct this negative effect.

The concept of "divergent thinking" describes well certain features of creative thinking. The use of divergent processes in describing creative production is especially justified with children because they are oriented toward play as a divergent activity, and primarily as a process, and not towards specific outcomes as is the case with convergent activities. In his model of human abilities, Guilford gives a detailed analysis of the disposition and function of divergent thinking and psychometrically tests their empirical existence. Guilford makes a distinction between adaptive and spontaneous expressions of divergent abilities. The first resemble convergent processes and are usually activated in the task of problem solving. The second are more unlike convergent processes and are usually activated in solving associative problems. The basic divergent problems, according to Guilford, are fluency, flexibility and originality, especially of ideas. Complexity, or rather, elaboration, is to a greater extent composed of convergent processes, even though it appears as a variable of divergent thinking.

THE HAT-DAT DIVERGENT CAPACITIES TEST

The associative divergent capacities test for children, named HAT-DAT is a modification of the Torrence Divergent Capacities Test for Children and Young People. The test measures the degree of expression of spontaneous fluency, flexibility and originality of ideas, as well as complexity or elaboration of ideas, although we did not deal with the last variable. Four of the nine subtests in the original Torrence version are used. The test is semistructured of the associative type. The children are individually tested in two sessions which are not strictly limited in duration. One session averages about 45 minutes and the interval between two sessions is a month and a half. Each time two problems are assigned, one figural and one verbal and they are assigned to all the subjects in the same manner and in the same order.

Subtest "Circles" – An A4 size piece of paper is set before the subject. In the upper left-hand corner, a small circle is drawn. The child is given colored pencils. The examiner says: "You see this circle? Now you try and draw here (he points along the edge of the paper) as many unusual round objects or things of which the circle makes up the main part. Do that as quickly as you can". The time for this assignment is not limited. The examiner interprets the meaning of the individual drawings, and writes their names underneath. The test is of the associative type and fluency, flexibility and originality of responses are scored.

Subtest "Kidney" – An A4 size piece of paper is set before the subject. Beside it is a piece of yellow, self-adhesive paper in the shape of a kidney, and the child is supposed to glue it on to the paper. The child is given colored pencils. The examiner says: "You see this yellow paper? First think of an unusual picture, one that no one else would think of, but remember, the main part of the picture is to be this yellow piece of paper. First imagine the picture in your mind, and once you do, decide where you need to stick the yellow paper, where exactly on this sheet of blank paper. When you stick it on, start drawing the picture you drew in your mind." The time for this assignment is not limited. Once completed, the examiner asks the child to give the picture a name, which is recorded. The test is not associative. It resembles a child's free production and is intended to measure only originality.

Subtest "Tin Can" – The child is given an open tin can without any decoration, (10 cm in diameter, 15 cm high). The examiner says:"Take a look at this tin can. Try to think of how it could be used in as many different and unusual ways as you can." The time for this assignment is not limited. Each response is recorded so that the smallest unit that represents a response can be distinguished and treated as a separate item. The test is of the associative type, and fluency, flexibility and originality are scored.

Subtest "Monkey" – The child is given a soft toy monkey, 40 by 10 cm in size, brown colored, without any decorations. The examiner says: "Look at this monkey. Try and think of as many unusual ways of making it more attractive." The examiner writes down everything the child says so as to discern the basic units to be scored. The test is of the associative type, and fluency, flexibility and originality are scored.

The measure of fluency was the number of answers given, not counting repetitions or those which contradicted the instructions. Flexibility was scored on the basis of several different criteria, (Torrence's, Kalmar's and ours), all of which basically amount to classifying each answer into a category of responses, in accordance with a given classification, and then counting the number of categories which the subject used while associating. The number of categories of answers used indicate the individual's inclination to change the direction of associations. Originality was also scored on the basis of several criteria, (Torrence's, Kalmar's and ours), all of which were based on determining the statistical frequency of an answer in the given sample of children. The least frequent answers were scored highest in originality. The sum of points for all responses represented the individual's score in originality on the given subtest.

Research began in 1978/79, with the enrollment of children into first grade elementary school (Josip Broz Tito, New Beograd), on a sample of 104 subjects of both sexes, and lasted over a period of four years, as a longitudinal study. The children were individually tested with the NBS Scale upon enrollment, when biographical data were gathered from the parents about their childrens' pre-school development. In first and third grade, the children were tested individually with the HAT-DAT (Beograd version of the Hungarian adaptation of the Torrence Divergent Capacities Test) test. In third grade, the children were tested in groups with the junior version of the modified (adapted) EPQ Personality Test, and examples of children's various works (free compositions, paintings, and free construction work) were collected and assessed by a number of specialists for creative content. In the fourth grade, the children were given the CPQ Test. We also collected data on general school achievement of each individual pupil, as well as achievement in relevant school subjects (Serbo-Croatian, art, technical education, and mathematics), during the period between first and fifth grade.

There were four phases to the data analysis: elaboration of the scoring criteria on the HAT-DAT Test, testing the psychometric characteristics of this test, selecting the best indicators of divergent abilities, and correlating them with the childrens' work products, intelligence, personality traits, school achievement and biographical data. The problems of scoring and psychometric characteristics of the HAT-DAT Test will be discussed here.

RESULTS

Developing criteria for scoring the HAT-DAT responses.

Besides using Torrence's and Kalmar's criteria in scoring the adapted test, we introduced additional criteria in order to achieve more precision in measuring certain variables. Our goal was to establish scoring methods which could best discriminate individuals in the sample.

1. Fluency, flexibility and originality of responses on the associative subtests was scored in two different ways. We considered all the answers that the subject gave, as well as only those given during the first 10 minutes. It is possible that a time limit frustrates the individual and fetters his originality which tends to appear in later phases of the association flow. However, we did not find significant variation in the results except in special cases of outstandingly fluent children.
2. Flexibility was scored using a special response classification, set up as a hierarchy of categories. The number of categories in the lower part of the hierarchy which the individual uses while associating, represents his flexibility score based on a mild criterion. The number of categories in the upper part of the hierarchy which the individual uses, represents his flexibility score on the basis of a more strict criterion. Shifting within the categories lower in the hierarchy is a much smaller step in the change of direction of thought, than are shifts higher in the hierarchy. Results showed a difference in flexibility when scored on different levels of strictness. Since the mildest flexibility scoring gives a score very close to fluency, and the strictest scoring (the categories from the top of the hierarchy) did not enable good discrimination between individuals in the sample, the categories of a medium level of generality seem the most adequate measure of flexibility.
3. Originality of responses was scored using the principle of statistical rarity of responses, but with an additional detail-elaboration and variation of scoring on different levels of strictness. The originality of a response was scored first through determining the statistical rarity of a given response in the sample of subjects. It was then scored on a percentage scale in which rare answers carry a maximum number of points. We also scored:
   a) originality of the first response in each narrow category, to avoid biases due to repetition of similar answers within the same category,

b) originality of responses leaving out those answers which belong to the high frequency category to minimize the influence of accidentally rare answers, and

c) originality of response categories on the basis of their frequency. We also find this criterion in Kalmar's scoring, but not separated from scoring of individual answers. However, we find very few rare categories of answers in the associative flow of individuals, so the discrimination of subjects based on this criterion is not satisfactory.

Psychometric Features of the Hat-Dat Test

1. The HAT-DAT Test is relatively subjective, despite the detailed explanation of the response scoring key. As a nonstructured test, with a goal of searching for rare responses, which are difficult to standardize while scoring, it leaves too much space for subjectivity of the evaluator in the case of repeated testing, where completely new, unscored and unclassified responses will appear. The responses which appear in our particular sample are all classified according to determined, clearly described percentage scales, or evaluated through classifications prior to scoring the responses of the subject. In other words, the same responses would be identified by the evaluator's using the key and through the availability of a score for each response from the list. The criteria were derived from Guilford's theoretical conceptions, applied later by Torrence, and later adapted by the Hungarian psychologists Magda Kalmar and associates. Even though there may be objections to this theoretical conception, endeavors were made that the criteria be defined and described as clearly as possible.

2. The test is discriminative and relatively difficult. On most of the indicators of all variables on all the test problems of the HAT-DAT Test, the subjects are well differentiated in achievement, except that the test sometimes (on certain indicators), shows to be of high standard, which can be noted on the basis of statistical evidence.

3. The test is unreliable in terms of internal consistency, i.e., uniformity and stability over time. The item correlations on both age levels are relatively low. This raises the issue whether viewing the variables as separate divergent capacities is justified, independent of the contents of the test

problem. Since the scoring variables within the test problem are highly correlated, they may represent one underlying variable. The problems thus acquire a status similar to compact parallel test forms which measure one variable. We may conclude that the test is not homogeneous, i.e., its internal reliability is relatively low.

The retest established that the HAT-DAT Test is unstable over time. Subjects do not retain the same rank in the sample on the same test indicators in first and third grade elementary school. Very few indicators correlate significantly with themselves on the retest or with other indicators. Even though there was a general improvement in the results, on most of the divergent thinking variables on the test between first and third grade, the unreliability of the test over time shows that the improvement does not rely on individuals evenly and in accordance with the results of the initial achievement. This means that divergence progresses with the age level with some children, that it stagnates with others, while it even decreases with some children. Reasons for the instability probably lie in the specific manner in which divergent abilities are manifested, which we assume depends to a great extent on the subjects' mood, and on individual variations in development.

4. The test is not valid in the sense of predictive validity and construct validity. Content validity was not tested for the reason that we wanted to test the value of the theoretical conception on which the test lies through correlating practical activity, as in free production. Results show that the divergent capacities test does not predict achievement in activities which may be assumed to be significant in the appearance of child creativity. A possible explanation could be that the test does not measure the same phenomenon. It only measures one aspect of divergence, which is only one of the conditions of creativity. Thus, divergence cannot in itself predict creative achievement. Another possible explanation may be the fact that the conception of creativity, based solely on the concept of divergent abilities and tests which measure it, is unsound.

Test Prediction and the Status of Divergent Abilities

We may comment on the status of different divergent abilities and the predictive value of the divergent thinking associative test. Considering the fact that this theoretical concept was based on statistics and derived from a test-like model of human abilities (Guilford's model), its statistical refutation represents its strongest criticism, a critique based on the arguments of the very system undergoing criticism, under the assumption that our data are sufficiently representative.

The problematic theoretical status of certain divergent abilities calls for reconsideration of this concept. The fact that it is not predictive, in the sense of test reliability, suggests that divergent abilities cannot be measured through statistical techniques, whereas, in the sense of test validity it indicates we have an inadequate content or level of generality in the concept of divergence, through which we identify the concept of creativity. The problems which arise in the process of establishing criteria in determining test validity indicate that even the non-testlike approaches to the assessment of child creativity cannot be considered to be reliable. Therefore, they are not satisfactory criteria, nor can they serve to test the value of the test as a measuring instrument to be used for the purpose of prediction.

The Medium

An increase of higher correlations between different variables of divergent thinking within the same test problem, and the absence of the same trend between achievements on the same variable on different problems, does not only question the status of divergent variables, but also implies that the medium through which the abilities are manifested plays a vital role. For example, there are no fluent subjects in general, but rather those who are fluent in certain test problems. However, results show that this situation is more strongly influenced by the contents of the problem itself and the current motivation of the subject than by the medium in which the problem appears. There is, for example, no essential link between indicators of fluency on two different verbal problems.

The Influence of Fluency on other Divergent Capacities in the Test

The achievements in flexibility and even originality, are influenced by the subject's fluency, because the correlation between these three variables, within the same test problem, are relatively high. Those subjects who show more fluency on the test, make a greater number of attempts, and therefore have a better chance to provide a more flexible and original response. For this reason, while scoring flexibility and originality, we need to somehow eliminate the effects of fluency on these variables. There are three possible ways to do this:

a) keep fluency constant by scoring flexibility and originality on the same number of responses for all subjects,
b) score flexibility and originality only in the initiative phase of the associative flow,
c) score flexibility and originality of responses by determining a percentage ratio between the number of these answers and the total responses.

The Concept of Transformation as an Indicator of Creativity

Even though fluency is least comprised of convergent factors, and is therefore the purest measure of divergence, it still gives us only information about the quantity of ideas, and not their quality. This fact may explain a growth in all divergent capacities (including originality) with age, despite our expectation that school may thwart children's originality, because originality on the test is influenced by fluency, which can be also stimulated by imitative school activities.

Creativity therefore, is a combination of divergence and convergence. If we wish to acquire a predictive instrument for measuring those capacities which are essential to creativity, then we need to establish the instrument on indicators of both convergent and divergent abilities. The concept of complexity, which combines divergent and convergent abilities and is based on transformation of the offered stimulus and thus comes closer to the situation of a spontaneous creative act, needs to be more closely investigated as a possible and more adequate measure of those abilities which are the basis of creativity. Even Guilford indicates something similar in his later works, in accordance with his theoretical model.

Gifted children: Research and education in France

Jean-Charles Terrassier
France

When Alfred Binet and Dr. Simon devised the first true intelligence scale at the turn of the century, their primary objective was to identify children who developed too slowly to keep up with normal teaching. Special classes, known as "improvement" ("perfectionnement") classes, set-up to meet their needs, are still being run today.

Binet was quick to realize that his test could also be used to identify gifted children. As early as 1909 he suggested that special classes be created for the gifted as well. As we know, the Binet test was adapted in the United States by Terman for use in his long-term investigation of the development of gifted children. This test, and various later versions, led to special classes being set-up in the USA, and a variety of educational approaches were based on it.

Meanwhile, in France, Binet's appeal fell on deaf ears. Even today, the French education system is a mass of contradictions. It actually forces gifted children to follow the same curriculum as other children at primary school and lower secondary school ("collège"). Over a period of ten years or so they are obliged to learn more slowly than their potential would allow. When they enter "lycée", sudden new demands are made on these gifted children. Far more is expected of them whereas, for years, they have been led to believe that school required no effort of concentration, industry, organisation or thought.

A wide-ranging French study (INED) which examined the intellectual level of children showed that 50% of gifted children (IQ of 125+) came from modest backgrounds (working-class parents). This subgroup of children with a more modest socio-economic background suffered a 50% failure rate in their secondary school studies despite their intellectual abilities. This does not compare favourably with the 90% success rate found among children from a better socio-cultural background.

These findings were not encouraging for those who favoured more social justice and equal opportunities. In addition, psychological and teaching science arguments presented in my book "Les enfants surdoués ou la précocité embarrassante" (Gifted children or embarrassing precocity) finally led the French Ministry of Education to ask me, in 1986, to develop a pilot scheme for special classes to meet the needs of intellectually precocious children. Our efforts to interest the Ministry of Education, begun in 1971, had taken 15 years to succeed in mounting a project which acknowledged the existence of these gifted children and their special needs.

This project, which I presented to the Ministry at the end of 1986, was quickly accepted and led to the creation of the first special class for intellectually precocious children at the Las Planas school in Nice in September 1987. The project's objective was to try to avoid the paradoxical situation where so many gifted children did not succeed academically in traditional schools. The project involved moderate acceleration, adapted to the rhythm of the children's development and taking into account their "dyssynchronies".

I introduced this notion of "dyssynchrony" at the World Conference on Gifted and Talented Children, in San Francisco, in 1977. It describes the heterogeneous development which is peculiar to those children whose precocity is limited to the intellectual domain. From my own observations I discovered that the emotional and psychomotor development of these children do not usually go hand in hand with their intellectual precocity. I dubbed this phenomenon "internal dyssynchrony" and it has important pedagogical implications. What I call "social dyssynchrony" is a result of the lag between the rate of normal development found in most children, and which has determined the learning rate in most schools, and the faster pace of the precocious child. Adults as well as children expect all children, precocious or not, to play a role that matches their age. This submits precocious children to what we call the "anti-pygmalion effect". They are encouraged to regress, levelling down to normal development through a process of self-rejection.

Every year since 1987 (with the exception of 1990) a new class for precocious children has been started at the Las Planas school in September. In 1990 the Académie Inspectorate of Nice curiously received only a very small number of admission files. Citing budgetary restrictions, the Académie Inspectorate also refused to create a fourth teaching post set aside for the fourth level of primary school, in accordance with the project.

The structure of the project had already been distorted through the unsatisfactory level of training among the teachers, through pressure to implement the project faster than planned, and through the scrapping of the Monitoring and Assessment Committee which was intended to oversee the progress of these classes. The teachers themselves, all highly motivated by their pupils, were often frustrated by the comings and goings of different hierarchical superiors (three different Académie Rectors and two académie Inspectors within a period of four years). As author of the project and expert appointed by the Minister, I could only protest against these harmful changes which undermined the confidence of teachers, children and parents alike.

These problems notwithstanding, the children made remarkable progress, undoubtedly more so and in a more balanced way than if they had remained in normal classes. However, the plan to monitor a control group of gifted children who remained in the normal schools had to be abandoned, thus reducing the validity of the assessment we were able to carry out.

Apparently the Education Minister has claimed that the latest primary school reforms would undoubtedly be an equally suitable means of meeting the needs of precocious children. Indeed, the right to diversity in education, which we at the ANPEIP have been promoting for some 20 years, has now been catered to in the latest reform put forward on 15 February 1990 by the Education Minister, Lionel Jospin. In a more flexible system precocious children might make progress by completing three levels in two years.

Of course, the spirit behind the latest reform is very encouraging. But, when stock is taken of the capacity of all educational systems for resisting change, and of the difficulty they have in breaking out of a routine, it is not unreasonable to fear that the application of this reform will not be easy and will have only patchy results. For this reason we believe it would be premature to think of abolishing the special classes at Las Planas until a valid assessment has been made of the real effects of the new school reforms on the educational success of precocious children.

Currently 2% to 2.5% of all pupils in France can benefit from skipping school-years, normally only one. However, precocious children whose parents belong to a high socio-cultural class (teaching, management, the professions), are more likely to have access to this option. This, in turn, helps to maintain the high failure rate among less socio-economically favoured categories of children.

The Ministry of Education can sometimes allow pupils to sit the Baccalauréat very early on, e.g., Arthur R., who passed it at the age of 11 years 10 months. However, depending on the specific teacher or inspector who is approached, parents may have to fight to have their child skip a school-year, even if they think it necessary. It is difficult to reconcile such disparities in treatment with equal opportunities.

In addition to the Las Planas state school, a private secondary school, the Lycée Michelet, in Nice, started a 1st-2nd-year class for precocious children in 1988. This was followed-up in 1989 with a 3rd-4th-year class and, in 1990, with a 5th-6th-year class. Usually, children admitted to the 1st-2nd-year class at Michelet are of the so-called normal age for entering secondary school because they were refused a chance to skip a school-year at primary school. The same admission criterion was applied as at the Las Planas state school, i.e. an IQ of 125+. At this stage in their education, some precocious children may achieve low marks as a result of an incipient lack of interest in schooling. However, poor marks do not prevent their admission to Michelet. The head-teacher, Michel Pinder, like us, has decided to take their intelligence into account instead of school performance. These children have not let us down. Most of them quickly regain interest in acquiring knowledge owing to their being in a stimulating environment away from the boredom of a slow learning rate and unnecessary repetition. Currently they end their 5th-year class at "lycée" after three years of study instead of five. Our objective is not to help them gain time, but simply to avoid their wasting it together with the feeling of boredom and lack of interest that would entail.

Apart from the ANPEIP, there are other initiatives in France aimed at offering extra-curricular activities. For example the "Jeunes Vocations" Club in Paris hosts rewarding activities on Wednesdays and Saturdays. The ALREP runs an annual summer camp offering a wide range of activities. The Institut Beaulieu, a psycho-therapeutic boarding establishment has for 20 years been welcoming a number of gifted children with emotional or learning difficulties, and conducting instructive research into what becomes of them. Researchers in the neuro-psychological field are currently assessing the possible link between intellectual precocity and electro-encephalographic indices. Theses and degree dissertations focussing on the gifted child are becoming less of a rarity.

However, considering this large number of initiatives, one can only conclude that, despite some signs of progress, gifted children are still neglected or even ignored by the French education system. It is, therefore, fortunate that the Council of Europe has undertaken a co-ordinating role in this area and thus will undoubtedly have the effect of stimulating more research.

Teaching gifted children and adolescents, and related research in Finland

Jouni Välijärvi
Finland

The Finnish school system has traditionally emphasized equality as an important educational aim. Particular attention has been devoted to the disadvantaged. During the last few years public debate has arisen on the special needs of gifted children and adolescents. Giftedness has on the one hand been seen as an individual quality that should be developed in a variety of ways, and on the other as a national resource that should be effectively utilized in developing society. The school system has been accused of levelling out individual differences and suppressing individuality. Even though recent legislation emphasizes that teaching has to be individualized to match the pupil's abilities, there have not been any significant changes in the teaching of the gifted. The debate, however, goes on, and concrete results are expected in the next few years.

The Finnish curriculum has never favoured individual choices. For example, in the upper secondary school there are 14 compulsory subjects for the pupil. Attitudes towards ability grouping of pupils are negative. In the comprehensive school permanent ability grouping (setting) has been abolished in all school subjects.

Catering to the special needs of the gifted is viewed as a task of the school and the teachers in the first instance. Teaching should be enriched, e.g. by using varied working methods and learning materials. Temporary special groups of the gifted and using other resource teachers are possible. Researchers have observed, however, that teachers do not see many opportunities to pay special attention to the needs of the gifted. Teachers are not able to perceive children's special talents nor provide guidance for their studies for the simple reason that they have received no training to be able to do this. It has been estimated that as many as half of the children in need of special arrangements are not identified.

Giftedness is sometimes manifested as behaviour problems at school, and the teacher may not be able to properly interprete what lies behind these problems. In terms of the development of a gifted child, it is important that his/her special talents are valued appropriately and that positive alternatives are available in the school environment for giftedness to manifest itself. Enriching teaching and increasing flexibility are, therefore, essential for developing giftedness.

The educational system in Finland allows a gifted pupil to skip a grade. In practise, however, this rarely occurs. Grade skipping can in any case be harmful if the child's intellectual and emotional maturation do not proceed in parallel, as is often the case with the gifted child.

Giftedness and its' manifestation in school has not been studied much in Finland. Uusikylä has studied the school experiences of gifted and talented children and their personality characteristics and hobbies. The group consisted of students with a high IQ (140 or higher). The gifted group included young people who were talented in mathematics, music, track and field, football and chess. They usually had positive experiences at school and the school success of the mathematically gifted was particularly good. They also reported that the school offered many opportunities for feeling successful and for making an effort. Giftedness had not normally hampered social development. However, boredom with school was fairly common and the gifted students were rather skeptical about the school's ability to foster independent thinking and creativity.

In comparing different gifted groups Uusikylä found that mathematicians are more conscientious than others, more introverted and particularly interested in technology. School anxiety was also fairly common and this seemed to be linked with the goal of obtaining perfect results. The musicians were sensitive and more unbalanced than the others and least satisfied with the school. Yet even their school success was excellent. The athletes were extraverts and popular in school, and were often leaders. The social background of musicians, mathematicians and chess wizards, in particular, was high and their homes favoured child-centered educational principles.

Uusikylä paid special attention to those who were underachievers at school. The IQ of the underachievers is high, but they do not do well at school. He noted that these students need special attention at school more than the others do. The teachers often had problems in recognizing the giftedness of these students and in treating them in an appropriate way. Such a youth is often reclusive, perhaps even aggressive, and his/her school experiences may vary considerably. They shared the following characteristics: they were more reflective than other adolescents, more submissive, more afraid of conflicts, and more sensitive. They were not as strong-willed, enthusiastic or ambitious as the other gifted youth.

Uusikylä collected his data by questionnaire. The positiveness of school experiences is perhaps overemphasized because the students being studied had already practically demonstrated their giftedness in many ways. This kind of study does not capture those young people whose special talents have remained unidentified at school. The essential question for a small nation like Finland is how much socially valuable potential talent is overlooked. In dealing with this question the school has two alternatives. It can develop its ability to measure talent and on the basis of this direct children and adolescents into special study programs. Some consider this desirable, even indispensable. Personally I would be in favour of another alternative. Specifically, the school should be sufficiently flexible in terms of organization and methods of work that every child and adolescent could develop as far as possible also in that particular field that he/she feels closest to heart.

Measures have been taken to make studying in the upper secondary school more flexible. In some schools the student can decide how many terms to spend on the upper secondary school studies. The student can graduate from this so-called gradeless upper secondary school in two years, whereas the ordinary upper secondary school takes three years. However, only 1 per cent of the students take advantage of this opportunity. My own studies show that the gradeless upper secondary school has clearly shifted responsibility for objectives, content and progress in studying to the students themselves. Gifted students, in particular, see this as a very positive development, since it introduces the desired flexibility into the upper secondary school. The efforts to increase elective studies in the upper secondary school also improve the gifted students' chances to develop their special talents. Whether these experiments become a permanent part of the Finnish school is still uncertain. In principle, all are in favour of increasing the share of elective subjects, but in practice a number of factors, e.g.,the contradictory viewpoints of different subject teachers, may slow down reforms.

Extra-curricular clubs are an important part of the school's activities. It is the clubs that offer the best opportunities for developing the students' talents at school. Experts are often hired to run the clubs. For instance artistic, expressive and sport ability is developed best thanks to the clubs.

There are no special schools for the gifted in Finland. However, some upper secondary schools have special lines for gifted students. The longest tradition in this direction is in educating the musically talented. There are also special classes for them in the compulsory comprehensive school. Nowadays several arrangements are being made for organizing the teaching of athletically gifted both in the upper secondary school and vocational schools. Teaching is normally arranged in the from of special lines in ordinary upper secondary schools. The schedule has been changed so that compulsory humanities subjects are partly replaced by their specialized studies and by training.

Some pilot projects focus on science and special attention is devoted to teaching methods and expanding teaching outside the walls of the school. In some schools students can specialize to some extent on foreign languages. These arrangements do not, however, differ much from the program of ordinary upper secondary schools. The teaching of the mathematically gifted is largely dependent on the teachers' ability to enrich the learning environment.

From the point of view of the balanced development of young people it is important to ask on whose terms giftedness is exploited. Especially for athletically talented students the interests of the school and of other interest groups sometimes appear to clash. As international competition is intensified it may prove very short-sighted to exploit the talents of children and adolescents. Within schools giftedness should be considered primarily from the student's point of view. This means that giftedness should be promoted as part of all-round personality development. Giftedness is a valuable resource which should never lead to "mental exploitation".

PART 4
Special Contributions

Managing education for talent development

John Raven[1]

Scotland

Studies conducted in the US, UK, and many other countries (Bill et al., 1974; Centre for Educational Sociology, 1977; De Landsheere, 1977; Johnston and Bachman, 1976; MacBeath et al., 1981; Morton-Williams et al., 1968, 1971; Raven, 1977) show that most parents, pupils, teachers, and employers know that post-primary "education" is a mess: it confers few benefits on students, does not deserve to be described as "academic" or "intellectual" (because there is little exercise of judgment, analysis or critical thinking), does little to help students to develop competencies which are useful at work or in society, and does little to help them to identify and develop their own particular high-level talents. Perhaps more seriously, the educational system almost completely fails to nurture the high-level competencies and areas of giftedness which are most urgently required to run our society effectively. If evidence for this last statement is needed one has only to look at the educational system itself (but see also Raven, 1986). There, one finds that, despite exceptions that are conspicuous because of their rarity, there is a distinct lack of the qualities which are needed to introduce and run an innovative and vibrant educational system which nurtures at least some of the most important talents of all of the children. This applies whether one looks at the politicians, evaluators, and public servants concerned, or at the teachers themselves. Parallel observations can be made in relation to the health service, agriculture, public order, defence, and the built or "natural" environment.

1 John Raven, Consultant on Education and Economic and Social Development, 30 Great King St., Edinburgh, EH3 6QH, Scotland

Most of the kinds of giftedness that are required to contribute effectively to organisations and society - including those needed at all levels in the public service - are relatively unrelated to academic giftedness (Gardner, 1990; Hatch and Gardner, 1990; Price, Taylor et al., 1971; Raven, 1984; Taylor, Smith & Ghiselin, 1963). Indeed, as Hope (1985), Nuttgens (1988), and Hogan (1990; Hogan et al. 1990) have demonstrated, children who are advanced into senior positions on account of their academic giftedness are not infrequently socially destructive of their organisations and society. What these observations mean is that educators need to be concerned with multiple talents and with identifying and developing the areas of giftedness of *all* children. Unfortunately, despite endless exhortation, few teachers do this. The neglect of these wider talents and the failure to introduce appropriate educational processes is not malicious, but due to a number of factors which have gradually emerged in the course of our research (see Raven, 1990, for a summary). To overcome these barriers it will be necessary to put in hand a wide range of research and development which is essentially psychological in nature - although much of it can only be carried out in the context of action-research and curriculum development activity within schools.

Public disquiet, based on direct observation of the muddle in education, has led directly to politicians imposing wave after wave of reforms onto the educational system over the past 40 years. None of these reforms have been grounded in an adequate understanding of the problem. As it has become clear that each was not working it has been followed, not by systematic enquiry into the causes of the problem and the reasons for the failure of the "solution" previously introduced, but by another "Great White Hope" based, yet again, in a superficial understanding of the nature of the problem. There have been few calls for research which would uncover the causes of the conspicuous malfunctions and the developments which are needed to overcome them. On the contrary, the very idea that research was *relevant* to the solution of such urgent and pressing problems has generally been treated with contempt.

In fact, while several studies (Bachman, et al. 1978; Flanagan, 1978; Goodlad, 1983; ORACLE; Raven, 1977) have demonstrated that there is indeed something very seriously wrong with post-primary education, the available evidence suggests that *what* is wrong is very different from what has commonly been assumed. Contrary to what Goodlad (1983) would have us believe, surveys of parents', teachers', pupils', and employers' opinions in the

US, UK, and other countries (Bachman et al., 1978; Bill et al., 1974; De Landsheere, 1977; Goodlad, 1983; Morton-Williams et al., 1968; Raven et al., 1985) show that the majority of each of these groups think that schools should be helping to foster qualities like initiative, the ability to work with others, problem solving ability, and the willingness to understand and influence society. Studies of the qualities required in the workplace and society support these opinions (see Raven 1984; 1990, for a summary). Yet teachers rarely employ the educational processes which would be required to foster such high-level talents in most of their pupils (Fraley, 1981; Goodlad, 1983; HMI, 1978; HMI (Scotland), 1980; ORACLE; Raven et al., 1985). A series of studies (Raven, 1980; 1984; 1990; 1991b) have enabled us to gradually build up an understanding of the reasons for this failure.

There are two main inter-related strands in the explanation. One has to do with the value-laden nature of the qualities that are to be fostered. The other involves society's failure to set the management of the educational system in the context of the understandings and structures that are required to promote innovation.

The value-laden base of the qualities which are to be fostered is immediately apparent if one reflects on the nature of the abilities that are needed to introduce change. These obviously include such things as the willingness and the abilities required to examine the way organisations and society work and identify leverage points at which change could be introduced, and the abilities needed to muster support for the necessary interventions. However, even such apparently innocuous goals as "enhancing cognitive development" turn out to be heavily value-laden: witness the furore which develops as soon as teachers set about encouraging children to ask questions or think for themselves.

The value-laden nature of competence in fact creates very serious dilemmas in connection with both the development and assessment of all high-level competencies (see Raven 1990; 1991b, for a fuller discussion). These dilemmas surface most forcefully in the shape of questions like whether one should give teachers power to 'brainwash' students and whether schools should assess value-laden motivational dispositions as part of the certification and placement process. Part of the problem is that, if we do not find ways of handling the dangerous powers we have unavoidably (by default if nothing else) given teachers in these areas, we render them worse than impotent: we actually force them into dysfunctional activities because we require them to suppress the development of important competencies in, and to convey inappropriate beliefs to, their pupils. Likewise, if one does not assess the values and political beliefs which are central to competent behaviour *as part of the certification and placement process* teachers and students will have to continue to neglect the development of such qualities because what happens in schools is *primarily* determined by what is assessed at the interface between schools and the wider society.

A number of surprising conclusions emerge from even these brief remarks. One is that, if we are to introduce an educational system which will identify and nurture high-level talents in *any* children, it will be necessary to generate a variety of distinctively different educational programmes, document the personal and social, short and long term, consequences of each, and feed that information to the public so that they can make their own choices about the type of programme in which they should enrol their children. Another is that, if the educational system is to identify and nurture a wide variety of very different talents and areas of giftedness, it will be necessary to (1) develop some tools to help teachers to identify each child's motives and incipient talents and create individualised developmental programmes in which the pupil can practice and develop high-level talents whilst carrying out activities which he or she cares about, and (2) to find some way in which assessments of these qualities can figure in the certification and placement process. We are therefore beginning to see the need for systemic - but not centrally prescribed system-wide - change if progress is to be made. The need for a major programme of innovative R&D is likewise beginning to become apparent, as is the need for new beliefs about the role of the public servant and the form of democracy required to handle choice. Finally, the dangers inherent in assessing value-laden talents alert us to the need to create an open system of public supervision so that malpractice can be avoided.

The second set of necessary developments indicated by our research are rather different. It emerges that, if the wide variety of talents needed by our society, and possessed by our children, are to be nurtured it will be necessary to create an educational system which is *permeated* by both a concern to invent better ways of meeting students' needs and develop a more effective educational system and the actual activities needed to do these things.

Many of the barriers to nurturing multiple high-level talents arise outside classrooms, e.g., from the expectations of parents, mandatory low-level-test--based "quality" control procedures (Raven, 1991c, discusses the seriously damaging consequences of employing inappropriate tests in research, policy evaluation, and personal evaluation, together with a discussion of the developments required to overcome the problems), and university selection practices. Other barriers arise from an absence of essential understandings of the nature and development of competence, an absence of appropriate measurement procedures, and an absence of tools to help teachers to administer individualised, competency-oriented, educational programs. Because of these barriers it will be necessary to:

(a) encourage teachers to work *outside* their classrooms to gain control over some of the wider sociological forces which deflect them from their goals, and

(b) establish a much more effective support structure. This support structure should include within it access to much more fundamental, but problem-driven, research carried out by researchers with the same innovatory concerns as are required among teachers and public servants and working within a similar structure to facilitate and encourage adventurous research which addresses value-laden and political issues.

Kanter (1985) has coined the term "parallel organisation activity" to refer to organisational climates pervaded by a concern with innovation accompanied by the kind of activity mentioned in the last paragraph and has described the expectations and arrangements that are needed in some detail (Raven, 1991b, has merged Kanter's work with other research on climates of innovation and spelt out the implications for the management of the educational system). It is, for example, essential to set aside time for the work, for all concerned to work in non-hierarchial relationships, for explicit emphasis to be placed on the importance of finding new things to do and new ways of doing things, to establish task forces having fluid memberships so that one person's talents can be complemented by those of others, and to channel resources to those who are best able to pull off adventurous activities rather than to those best able to generate paper plans.

The developments needed in the educational system are so numerous, varied, and pervasive that they cannot be introduced by central prescription, or monitored by centralised testing of the kind that has been widely advocated over the past decade. (The futility of central prescription of goals and procedures is even immediately apparent if one reflects on the need for teachers to exercise a great deal of discretion and creativity if they are to nurture the very different talents of the children in their classes).

If what has been said is correct, new arrangements are required to manage provision. An example will illustrate the point. If the necessary variety and choice of educational programmes is to be provided, it will be necessary for the public service to generate a range of options and carry out (or commission) the research needed to document the short and long term consequences of each option for the pupils concerned and for the societies in which they live - and then provide those alternative programmes in every community. The public service will also have to produce evidence that each of the options is of high quality and then publicise that information (as well as information on the consequences of each option) so that pupils and parents can make informed choices between them. Finally, it will have to identify and remove deterrents to pupils enrolling in particular programmes. If all this is to happen, public servants will have to embrace a much more pro-active, innovation-oriented, and informative role than they have tended to assume in the past.

It is particularly important to note that such arrangements would require public servants of all kinds - from teachers to superintendents - to feed information outward to the public instead of upward in a bureaucratic hierarchy to elected representatives. At one level this would allow students and parents to make many decisions for themselves instead of having courses of action imposed on them by bureaucrats or elected representatives. But, at a more basic level, such arrangements would amount to new (participative) forms of democracy.

If public servants are to play this major role in running society, new arrangements will be needed to supervise their work. Otherwise the government overload which currently leads to such ill-informed and ill-considered decisions will only get worse. It is the public servants who should be taking decisions which are in the long term public interest: the public's task is to ensure that they do so. This is to be achieved by finding more effective ways of exposing the work of public servants to the public gaze. To do this, much more use will have to be made of network-based supervisory structures (involving the public and the media) of the kind discussed by Schon (1987), Toffler (1980) and Ferguson (1980). As far as education is concerned, these supervisory groups will need to have fluid but overlapping memberships involving parents, teachers, other members of the community, administrators and the media, and they will need to monitor the work of individual teachers, schools, groups of schools, local authorities and international research. The work of these supervisory groups will need to be supported by much better formal staff- and organizational- appraisal data collected by professional evaluators or using professionally-developed tools. The relevant information will need to include such things as data on individuals' capacities to seek out and act on data in an innovative way in the long term public interest and their capacity to contribute in one or other way to an innovative climate dedicated to finding better ways of doing things - or what Gardner (1990) has called "climates of intelligence". It will also need to include information on such things as what is happening in particular classrooms and what the long term personal and social consequences of alternative actions are likely to be. Such data can only be provided through the development and use of more appropriate staff appraisal tools and the conduct of more professional organisational- and policy appraisals. And it will only be used if more effective open-government structures are developed. These are discussed in slightly more detail in a number of previous publications (Raven 1983, 1989a,b; and Raven, 1991a,b, in press).

In summary, managing the educational system in such a way as to nurture the diverse talents of different children - diverse talents that are urgently needed to create the climates of socially-responsible innovation and intelligence that are crucially required to avoid the collapse of our society - therefore requires a generally unanticipated set of developments:

- it involves acknowledging the crucial role which our public servants play in the management of society;
- it involves creating a situation in which those public servants deliberately set out to establish "parallel organisation" activity to promote innovation;
- it requires those public servants to much more systematically collect evaluation data which examines the operation of social systems *qua* systems, sift that information for good ideas, and act on it in an innovative way in the public interest;

- it involves staff-appraisal systems which will enable public servants to get credit for engaging in such activity;
- it involves establishing forms of public supervision which both induce public servants to act in the long term public interest and provide the public with the information they need to assure them that the powers they have given their public servants are not being misused;
- it involves requiring public servants to generate variety, amass information on the consequences of each option, and feed that information to the public instead of upward in a bureaucratic hierarchy to elected representatives - i.e. it involves the introduction of new (participative) forms of democracy;
- it involves the initiation of the research needed to develop a better understanding of the way society works and the developments needed to make it work better (i.e. it involves the application of the concepts, tools, and methods of organisational psychology to the management of society as an organisation);
- it involves the initiation of the research needed to develop a much better understanding of the nature of competence, the educational processes that are to be employed to nurture its components, and the tools that are needed to orchestrate multiple high-level developmental programmes and to evolve the societal arrangements that are needed to ensure that all of these talents will be utilised and rewarded (i.e. it involves establishing many more policy evaluation, research, and development units, managing those units in new ways, and holding those who work in them accountable against new criteria).

It would therefore seem that there is no shortage of points at which gifted, creative, and socially-committed people can take action.

However it would not be appropriate to conclude this paper without noting that constructive developments in these areas are not only essential to the introduction of an *educational* system which nurtures and recognises the crucially important gifts of our children: they are also crucial to the very future of our society. They are the arrangements we need to evolve the developments that are needed to find ways of running our society that are much less destructive than those we have at present - i.e. they are the arrangements that are needed to find ways of countering the environmental destruction and pollution, over-consumption of renewable and non-renewable resources, destructive "defence" systems, inequitable trade, and mis-management of the financial system which threaten the very future of the world as we know it. They are the developments we need to give effect to the "new values" which so many of our citizens now embrace. Ironically, effective programmes of civics education in schools would be centrally concerned with promoting the evolution of these new understandings, tools and structures. Yet, in a "catch 22" situation, the introduction of effective multiple-talent educational programmes is dependent on the introduction of the understandings, mechanisms, and structures which it would be the objective of the programme to evolve.

And so we come full circle. Generating the requisite understandings and procedures will demand contributions from a wide range of people who possess very different high-level talents. Yet, conspicuously, the often academically outstanding people we have promoted into senior management positions in our society have not only not even begun to analyse and tackle the problems which beset our society, they often appear to be pre-occupied with their own advancement in ways that are not easily reconciled with a commitment to seek out and act on information in the public interest. In our educational system we need to nurture, and promote, people who possess many different types of giftedness. Developing the understandings and inventing the tools and arrangements that are required to do this is *the* central problem of the educational system at the present time. It is therefore on bringing these developments into being that those concerned with the education of gifted children need to focus.

References

Bachman, J.G., O'Malley, P.M. and Johnston, J. (1978). *Adolescence to Adulthood: Change and Stability in the Lives of Young Men.* Ann Arbor, Michigan: The Institute for Social Research.

Bill, J.M., Trew, C.J. and Wilson, J.A. (1974). *Early Leaving in Northern Ireland.* Belfast: Northern Ireland Council for Educational Research.

Centre for Educational Sociology, University of Edinburgh. (1977). *Collaborative Research Dictionary.*

De Landsheere, V. (1977). On defining educational objectives. *Evaluation in Education, 1, No.2.* Oxford: Pergamon Press.

Ferguson, M. (1980). *The Aquarian Conspiracy.* London: Paladin.

Flanagan, J.C. (1978). *Perspectives on Improving Education from a Study of 10,000 30-year-olds*. New York: Praeger Publishers.

Fraley, A. (1981). *Schooling and Innovation: The Rhetoric and the Reality*. New York: Tyler Gibson.

Galton, M. and Simon, B. (1980). *Progress and Performance in the Primary Classroom*. London: Routledge and Kegan Paul.

Galton, M., Simon, B. and Croll, P. (1980). *Inside the Primary Classroom*. London: Routledge and Kegan Paul.

Gardner, H. (1990). The difficulties of school: Probable causes, possible cures. *In Literacy in America, DAEDALUS/ Proceedings of the American Academy of Arts and Sciences, 119*, 85-113.

Goodlad, J. (1983). *A Place Called School*. New York: McGraw-Hill.

Hatch, T.C. and Gardner, H. (1990). If Binet had looked beyond the classroom: The assessment of multiple intelligences. *International Journal of Educational Research*, 415-429.

HMI. (1978). *Primary Education in England: A Survey by H. M. Inspectors of Schools*. London: Dept. Education and Science: HMSO.

HMI. (Scotland) (1980). *Learning and Teaching in Primary 4 and Primary 7*. Edinburgh: HMSO.

Hogan, R. (1990). Unmasking incompetent managers. *Insight, May 21*, 42-44.

Hogan, R., Raskin, R. and Fazzini, D. (1990). *The Dark Side of Charisma*. Tulsa: Institute of Behavioural Sciences.

Hope, K. (1985). *As Others See Us: Schooling and Social Mobility in Scotland and the United States*. New York: Cambridge University Press.

Johnston, L.D. and Bachman, J.G. (1976). Educational Institutions. *in* J.F. Adams (Ed.), *Understanding Adolescence,* IIIrd Edition, (pp 290-315). Boston: Allyn and Bacon.

Kanter, R.M. (1985). *The Change Masters: Corporate Entrepreneurs at Work*. Hemel Hempstead: Unwin Paperbacks.

MacBeath, J., Mearns, D., Thomson, B. and How, S. (1981). *Social Education: The Scottish Approach*. Glasgow: Jordanhill College of Education.

Morton-Williams, R., Finch, S., Poll, C., Raven, J. and Ritchie, J. (1968). *Young School Leavers*. London: HMSO.

Morton-Williams R., Raven, J. and Ritchie, J. (1971). *Sixth Form Teachers and Pupils*. London: Schools Council/Books for Schools.

Nuttgens, P. (1988). *What Should We Teach and How Should We Teach It?* Aldershot: Wildwood House.

ORACLE. See Galton & Simon (1980), Galton, Simon & Croll (1980), Simon & Willcocks (1981).

Price, P.B., Taylor, C.W., Nelson, D.E. et al. (1971). *Measurement and Predictors of Physician Performance: Two Decades of Intermittently Sustained Research.* Salt Lake City: University of Utah, Dept. of Psychology.

Raven, J. (1977). *Education, Values and Society: The Objectives of Education and the Nature and Development of Competence.* Oxford: Oxford Psychologists Press.

Raven, J. (1980). *Parents, Teachers and Children: An Evaluation of an Educational Home visiting programme.* Edinburgh: The Scottish Council for Research in Education.

Raven, J. (1983). Toward new concepts and institutions in modern society. *Universities Quarterly, 37,* 100-118.

Raven, J. (1984). *Competence in Modern Society: Its Identification, Development and Release.* Oxford: Oxford Psychologists Press.

Raven, J. (1986). A nation really at risk: A review of Goodlad's "A Place Called School". *Higher Education Review, 18,* 65-79.

Raven, J. (1989a). Democracy, bureaucracy and the psychologist. *The Psychologist, Vol.2, No.11,* 458-466.

Raven, J. (1989b). Equity in diversity: The problem posed by values and their resolution. *in* F. MacLeod (Ed.), *Families and Schools: Issues in Accountability and Parent Power.* Brighton, England: Flamer Press.

Raven, J. (1990). The barriers to achieving the wider goals of general education. *The British Educational Research Journal, 16, No.3,* 273-296.

Raven, J. (1991a, in press). *Giving Effect to the New Values.* New York: Trillium Press; Oxford: Oxford Psychologists Press.

Raven, J. (1991b, in press). *Managing the Educational System for Effective Schooling: The Most Important Problem is to Come to Terms with Values.* New York: Trillium Press; Oxford: Oxford Psychologists Press.

Raven, J. (1991c). *The Tragic Illusion: Educational Testing.* New York: Trillium Press; Oxford: Oxford Psychologists Press..

Raven, J., Johnstone, J. and Varley, T. (1985). *Opening the Primary Classroom.* Edinburgh: The Scottish Council for Research in Education.

Schon, D. (1987). *Educating the Reflective Practitioner.* San Francisco: Jossey-Bass.

Simon, B. and Willcocks, J. (Eds.). (1981). *Research and Practice in the Primary Classroom.* London: Routledge and Kegan Paul.

Taylor, C.W., Smith, W.R. and Ghiselin, B. (1963). The creative and other contributions of one sample of research scientists. *in* C.W. Taylor and F. Barron (Eds.), *Scientific Creativity: Its Recognition and Development.* New York: Wiley.

Toffler, A. (1980). *The Third Wave.* New York: Bantam Books.

Literature on giftedness - Where to search, what to expect

Ernst A. Hany[1] *& Gerhard W. Hany*
Germany

If one wants to gain an overview of activity in giftedness research and nurturing in Europe, one has recourse not only to regionally recognised experts but also to written documents. This contribution analyzes a number of information sources and gives advice on obtaining literature and on the effective design of literature reviews. Problems of content and methods which arise in connection with the design of literature surveys will, however, not be dealt with here. Clinkenbeard (1991) gives useful advice on this and, in addition, recommends the accounts given by Cooper (1984), Light & Pillemer (1984), and Borg & Gall (1989). Details of recent review articles and of other publications on giftedness are given in the appendices to the present contribution.

HOW TO FIND LITERATURE ON GIFTEDNESS

The most important sources of information on giftedness research are scientific books and journals. Surveys of the state of the art and results of current research are presented at conferences and published in conference proceedings. A large number of these documents are indexed in literature data banks. These sources will now be discussed in more detail.

---

1 Dr. Ernst Hany, Ludwig-Maximilians-Universität, Institut für Empirische Pädagogik und Pädagogische Psychologie, Abteilung Psychologische Diagnostik, Leopoldstraße 13, D-8000 München 40

Journals

The particularly rich tradition of giftedness research in the USA has lead to the founding of a number of relevant journals which are indispensable for literature searches. Table 1 shows the journals with the most contributions on the topic of giftedness research in the years 1986-1991 on the basis of a literature search in the Psychological Abstracts. (The details of this literature search are given below.) The most important journals are these three:

- Gifted Child Quarterly,
- Journal for the Education of the Gifted,
- Roeper Review.

Table 1

The American journals with the most frequent contributions on giftedness (Source: PsycLit 1986-1991; descriptor "GIFTED")[2]

	1986	1987	1988	1989	1990	1991	total
Creative Child and Adult Quarterly	4	10					14
Gifted Child Quarterly	36	29	21	21	19		126
Gifted Educational International	17	5					22
Journal for the Education of the Gifted	19	23	12	18	16		88
Journal of Counseling and Development	16	1	2	1	1		21
Journal of Educational Psychology		1	2	1	10		14
Psychological Reports	3	5	1	2	4	2	17
Psychologie in Erziehung und Unterricht	1	2	3	3	3	2	14
Psychology in the Schools	9	1	3	6	3		22
Roeper Review	54	30	35				119

[2] The volumes for 1990 and 1991 may not have been fully indexed at the time of the authors' date base access (December 1991)

It is pleasing to note that, as pointed out by Mönks (1990), a European journal is to be found among those with the most frequent contributions: *Psychologie in Erziehung und Unterricht* (Psychology in Education and Instruction), edited by Horst Nickel and Kurt Heller. This journal contained in most of the years investigated even more relevant contributions than the *Journal of Counseling and Development* or the *Journal of Educational Psychology*, - these Journals only gained a place in Table 1 because in the years covered, each dedicated one entire issue to the topic of giftedness.

Not included in Table 1, because it is not indexed in the data bases, but of increasing importance especially for European researchers is the *European Journal of High Ability*, founded by the European Council for High Ability and now, in 1992, in its third year of issue. It is however striking that European researchers seldom publish articles in the big American journals. Interesting information special to local nurturing programmes or on political developments in the area of giftedness promotion can better be obtained from newsletters, which have however only limited scientific aims or contents.

Literature data banks, discussed below, give information on the contents of numerous journals. Searches were made in these data banks for recent summaries of work on giftedness. Articles were identified from 1986 onwards which bore the descriptor "gifted" and which were either specified as a "bibliography" or "literature review" or for which it was clear from the abstract that they contained a literature review. A short selection from the results of this search is provided at the end of this chapter following the references.

Information can also be obtained from printed bibliographies on selected giftedness topics. These include the bibliographies published by ERIC, the information centre of the *American Council for Exceptional Children* through its *Clearing House on Handicapped and Gifted Children*, and the reviews of recent articles in various journals on particular topics, published from time to time in the *Journal for the Education of the Gifted.* The following themes have been covered recently:

Volume 14, No. 1, pp. 101ff.:	Co-operative learning (12 articles)
Volume 14, No. 3, pp. 312ff.:	Gifted underachievers (8 articles)
Volume 14, No. 3, pp. 317ff.:	Culturally different gifted (12 articles)

Volume 14, No. 3, pp. 321ff.: Gifted students with disabilities (10 articles)
Volume 14, No. 3, pp. 325ff.: Young gifted (5 articles)
Volume 14, No. 3, pp. 331ff.: Gifted females (9 articles).

Some other journals also review recent publications from time to time (e.g., Geffen (1989) summarises 12 dissertations). If one wants to read particularly those articles that were read and quoted by many other researchers, one can be referred to Carter & Swanson (1990) who have collected together those (American) articles which were quoted most often in other work.

Books

Solid information on recent monographs and anthologies can be obtained from

- book reviews in journals on giftedness,
- book reviews in the journal *Contemporary Psychology*,
- book details given in the data bank PsycBOOKS (Granick, 1990; Walker, 1991).

Information on those sources and a list of newer book publications are available from the authors.

Congresses and Conference Proceedings

The best way to follow current research projects, find out which research areas are most in vogue, and to make contact with like-minded researchers is to visit international congresses on giftedness and to read the accompanying congress proceedings. The Proceedings of the Ninth World Conference on Gifted and Talented Children, which was held in The Hague in 1991, have recently been published (Mönks, Peters & George, 1992). The titles of the Proceedings of all conferences organised by the World Council for Gifted and Talented Children (WCGT) together with a list of the adresses of the secretariates of the conferences of the European Council for High Ability (ECHA) can be obtained from the authors. Unfortunately, European authors have been relatively rarely represented in the WCGT congress volumes, apart from those occasions on which the congress took place in Europe (see Heller & Menacher's (1991) analysis of those congress volumes).

Literature data banks

A number of institutes have been created in order to provide academics with rapid overviews of information media. A number of data bases are relevant to giftedness research. Two of these, for which personal computer versions are also available, will be discussed below. In addition to these, three other data banks deserve a brief mention. The first of these is the *Exceptional Child Education Resources*, published as a quarterly journal by the *Council for Exceptional Children*. This data bank gives information on journal articles, books, congress contributions, manuscripts and dissertations. Of particular interest is the fact that the entries for most of the documents specify how they can be obtained (e.g., through international photocopying and microfiche services). In a comparative study of data banks, Rogers (1989) found a great deal of general literature on giftedness in this data bank. The second data bank, also appearing in text form, is the *Child Development Abstracts and Bibliography*, published three times a year by the University of Chicago Press for the *Society for Research in Child Development*. As well as thc usual information, this data bank contains the full correspondence addresses of the authors (for questions or for ordering reprints). Finally, further national and international data banks on pedagogical research in general are described in the *Bulletin of the International Bureau of Education* (vol. 63, Nr. 251, issue April-June 1989).

The following data bases will now be discussed in more detail: The American Psychological Association produces the *Psychological Abstracts* (see Auld, 1991), in which articles from 1300 journals from 50 countries in a total of 27 languages are summarised in English. This data bank is available in a computer-supported form, PsycINFO, which in conjunction with the software PsycLIT enables individual literature searches to be carried out on a personal computer with a CompactDisk player.

A still more extensive data bank is produced by the *Educational Resource Information Center* (ERIC). The computer supported version integrates two information sources: *Resources in Education* (RIE; comprises project presentations, reports of practical work, teaching materials, research papers, tests and other documents) and the *Current Index to Journals in Education* (CIJE; based on academic journals). The data bank includes over 600,000 documents to date (Barnett, 1987).

Evaluation of these data bases

There now follows a discussion of how far both of these data banks fulfill two criteria, which seem to us to be essential for effective literature searches on giftedness:

(1) Comprehensiveness of the entries, in the sense that as many information sources as possible (journals etc.) should be as comprehensively processed as possible (i.e., that all years are covered);

(2) Degree of detail in the document analysis, in the sense that the most important concepts of giftedness research should be available to control the data search.

Both data bases offer essentially the same information; i.e., for each document author details, title, year of publication, further bibliographical details, language of publication, keywords for searches and an abstract. Due to the heterogenous nature of the documents included, ERIC offers additional information on target group (e.g. researchers or practitioners) as well as three levels of keywords (major descriptors, minor descriptors, indicators), whereas the PsycLit dataset includes for each document not only descriptors but also special concepts, assisting detailed searches. In a trial search of all literature since 1986 we found that the ERIC data bank contains more documents on giftedness than the PsycLit data bank and that a full-text search of key words leads to more "hits" than a search which is restricted to descriptors.

Both data bases have a significant number of gaps. Table 1, which analyses the PsycLit databank, shows that the most recent years of the journal *Gifted Education International* and *Roeper Review* do not seem to have been processed. As a result of a systematic investigation, we found that there are also discrepancies between the data bases with respect to other journals. These discrepancies could be due to:

- the processing beginning or ending at differing volume and issue numbers,
- new issues being processed at different speeds,
- differing selection criteria for journal articles (e.g. conference reports, editorials, book reviews).

To date (January, 1992) neither data bank has included the *European Journal for High Ability*. In summary, it must be said that the user must pay special attention to the information in the data bank manuals which specifies which information sources were covered in which period of time.

Considered from a European perspective, it is striking that contributions in non-English languages are rarely represented, especially in ERIC, which restricts itself almost exclusively to English (frequencies are not shown here). PsycLit contains a surprising number of entries in German, whereas other languages - at least in the selected documents on giftedness - are only marginally represented. An additional analysis was carried out on how frequently entries on European countries appear in the selected contributions on giftedness (i.e., 746 from PsycLit and 1806 from ERIC) and compared with a search in the German DIPF data base. The results are distressing for European giftedness research. PsycLit documented 178 occurrences from 12 countries; ERIC documented 46 from 11 countries; DIPF documented 1605 occurrences from 23 countries. It seems that the two big American data bases document this activity inadequately.

These results justify the conclusion that the criterion "comprehensiveness" is, from a European point of view, not fulfilled by either of the data banks. Sadly, the situation is not very different with the criterion "degree of detail". In order to evaluate this criterion, the descriptors most frequently used for articles on giftedness have been considered. Each data bank takes its descriptors from its own thesaurus (Houston, 1987; Walker, 1991). Many of the most frequent descriptors are used to designate particular age groups and school types. The other concepts are, however, mostly kept very general. The reason for this is that in the data banks' thesauruses (that is, the hierarchically ordered descriptor catalogues), there are only a few special descriptors for identifying giftedness research. For this reason, PsycLit has to, for example, reduce all forms of diagnostics in giftedness research (identification, project impact, measurement of change, evaluation) to the descriptor "educational diagnosis". This means that in neither data bank it is easy to identify special areas of giftedness research only by using the descriptors. One must make recourse to full text searches or to additional keywords (PsycLit: key terms; ERIC: indicators), which are not however always systematically or comprehensively entered in the documents.

A EUROPEAN DATA BANK ON GIFTEDNESS LITERATURE

It has been recognised for many years that the existing American data bases do not give sufficient information on European giftedness research. With this in mind, work was begun in Frankfurt, West Germany, on a project led by Hansgeorg Bartenwerfer which had the aim of setting up a documentation system for collecting and indexing papers on giftedness. However the funding, from the German Ministry of Education and Science, ended after three years, with the result that although it had been possible to compile a data bank comprising more than 5000 titles, preparations for publication were left incomplete. Two volumes addressed to the German-speaking public were completed (Bartenwerfer & Müller, 1985; Bartenwerfer, 1988) before the project was shelved for the time being in 1987.

This data bank, which was prepared at the German Institute for International Educational Research (DIPF), has special features. It contains contributions from a large number of European countries. A large range of languages is also included. The special nature of giftedness terminology is reflected in the comprehensive system of descriptors - for instance, domain-specific descriptors are provided. To assist international use, all non-English titles were translated into English; and for many entries there are abstracts in English. Older titles from the nineteen-fifties are also included, making the data bank of especial interest for historical research.

Unfortunately the data bank also has its weak points. In addition to numerous typographical errors, the lack of documentation on the indexing procedure is to be lamented; there is no way of knowing which sources were processed and how thoroughly or regularly this was carried out, nor of knowing which rules were followed by the assignment of descriptors or what form the Thesaurus took.

In 1990, the German Ministry of Education and Science made new funds available to bring this data bank up to date and to make its contents available to the public. At the University of Munich (Germany), work has begun on the revision and completion of the existing material, under the leadership of Kurt Heller. An up-to-date overview for the German reader should be available by 1992. Work on widening the availability to an international readership is still in progress. The project involves the development of custom software based on the data bank system PARADOX.

The aims of this international literature documentation are as follows:

(1) to collect and index documents on giftedness from the whole of Europe and neighbouring countries;

(2) to identify gaps in the existing international data banks on European giftedness research and nurturing;

(3) to pass the assembled information on to interested researchers, teachers, educationalists and politicians.

A major step towards achieving these aims would be made by the establishment of a European network for the processing and exchange of material. This would require the cooperation of European research institutes, organisations for the promotion of giftedness, congress organisers and journals editors and publishers. Such a data bank would not be limited to scientific work; working papers, research reports, legal texts, short descriptions of promotional measures or developments in educational politics would also be important. Alongside bibliographical details, other information on the author would also be stored, following their permission, facilitating communication between people and institutions throughout Europe. A targetted search using the specially constructed thesaurus would identify not only relevant documents but also contact addresses, which could be passed on to approved individuals, with due provision for data protection. Data would either be entered by the team at the University of Munich on the basis of available literature, or by others cooperating in the scheme throughout Europe, using specially prepared forms. If financial support for the establishment of a European network is indeed made available by a European Community fund, information could be exchanged by international electronic mail.

References

Auld, D. (Ed.). (1991). *Psychological Abstracts. Nonevaluative summaries of the serial literature in psychology and related disciplines.* Arlington, VA: APA.

Barnett, L. (1987). Indexing and retrieval in ERIC: The 20th year. In J. E. Houston (Ed.), *Thesaurus of ERIC Descriptors.* (11th ed.) (pp. X-XXVI). Phoenix, Arizona: Oryx Press.

Bartenwerfer, H. (1988). *Bibliographie zur Hochbegabung* (2. Auflage). [Bibliography on Giftedness (2nd ed.).] Köln: Nomos.

Bartenwerfer, H. & Müller, H. (1985). *Bibliographie Hochbegabung. Deutschsprachige Literatur.* [Bibliography on Giftedness. German Language Literature.] Frankfurt am Main: Deutsches Institut für Internationale Pädagogische Forschung (DIPF).

Borg, W. R. & Gall, M. D. (1989). *Educational research: An introduction* (5th ed.). New York: Longman.

Carter, K. R. & Swanson, H. L. (1990). An analysis of the most frequently cited gifted journal articles since the Marland Report: Implications for researchers. *Gifted Child Quarterly, 34,* 116-123.

Clinkenbeard, P.R. (1991). Beyond summary: Constructing a review of the literature. In N. K. Buchanan & J. F. Feldhusen (Eds.), *Conducting research and evaluation in gifted education* (pp. 33-50). New York: Teachers College Press.

Cooper, H. M. (1984). *The integrative research review: A systematic approach.* Beverly Hills, CA: SAGE.

Geffen, L. (1989). Recent doctoral dissertation research on gifted. *Roeper Review, 12,* 122-124.

Granick, L. (Ed.). (1990). *PsycBOOKS 1989. Books & Chapters in Psychology, Vol. 1 to 5.* Arlington, VA: APA.

Heller, K.A. & Menacher, P. (1991). State of the art of research on giftedness and talent in the Proceedings to the WCGT Conferences since 1975. *Paper presented at the 9th World Conference on Gifted and Talented Children,* The Hague, July 1991.

Houston, J. E. (Ed.). (1987). *Thesaurus of ERIC Descriptors* (11th ed.). Phoenix, Arizona: Oryx Press.

Light, R. J. & Pillemer, D. B. (1984). *Summing up: The science of reviewing research.* Cambridge, MA: Harvard University Press.

Mönks, F. J. (1990). Hochbegabtenförderung als Aufgabe der Pädagogischen Psychologie. [Gifted education as a task of educational psychology.] *Psychologie in Erziehung und Unterricht, 37,* 243-250.

Mönks, F. J., Peters, W. A. M. & George, D. (Eds.). (1992). *Talent for the future: Social and personality development of gifted children.* Assen, the Netherlands: Van Gorcum.

Rogers, K. B. (1989). A content analysis of the literature on giftedness. *Journal for the Education of the Gifted, 13,* 78- 88.

Walker, A. jr. (Ed.). (1991). *Thesaurus of psychological index terms* (6th ed.). Arlington, VA: APA.

Selected review articles

Gockenbach, L. B. (1989). A review of personality factors in parents of gifted children and their families: Implications for research. *Journal of Clinical Psychology, 45,* 210-213.

Gonzalez, J. & Hayes, A. (1988). Psychosocial aspects of the development of gifted underachievers: Review and implications. *Exceptional Child, 35,* 39-51.

Halpern, M. (1986). Sources of Information on Tests for Special Populations. *Education Libraries, 11*, 73-76.

Hampson, R. B. (1988). Special foster care for exceptional children: A review of programs and policies. *Children and Youth Services Review, 10*, 19-41.

Keirouz, K. S. (1990). Concerns of parents of gifted children: A research review. *Gifted Child Quarterly, 34*, 56-63.

Kissinger, P. (1986). Recent Reference Sources on the Gifted. *Education Libraries, 11*, 82-86.

Olszewski, P., Kulieke, M. J. & Buescher, T. (1987). The influence of the family environment on the development of talent: A literature review. *Journal for the Education of the Gifted, 11*, 6-28.

Olszewski-Kubilius, P. M., Kulieke, M. J. & Krasney, N. (1988). Personality dimensions of gifted adolescents: A review of the empirical literature. *Gifted Child Quarterly, 32,* 347-352.

Rogers, K. B. (1986). Do the gifted think and learn differently? A review of recent research and its implications for instruction. *Journal for the Education of the Gifted, 10*, 17- 39.

Sowell, E. J., Bergwall, L. K., Zeigler, A. J. & Cartwright, R. M. (1990). Identification and description of mathematically gifted students: A review of empirical research. *Gifted Child Quarterly, 34*, 147-154.

Traxler, M. A. (1987). Gifted education program evaluation: A national review. *Journal for the Education of the Gifted, 10,* 107-113.

LIST OF PARTICIPANTS

1. Chairman, Rapporteur and Lecturers

Prof. Dr. Franz MÖNKS, (Chairman) Director
Center for the Study of Giftedness
PO Box 9104
NL-6500 HE NIJMEGEN

Drs. Herman W. VAN BOXTEL (Rapporteur General)
Faculteit der Sociale Wetenschappen
Rijksuniversiteit Utrecht
Postbus 80140
NL-3508 TC UTRECHT

Prof. Dr. Åke EDFELDT
Institute of Education
University of Stockholm
Frescati Hagväg 24
S-106 91 STOCKHOLM

Dr. Joan FREEMAN
University of London
Institute of Education
21 Montagu Square
GB-LONDON WIH IRE

Prof. Dr. Candido GENOVARD (excused)
Universidad Autónoma de Barcelona
Aptdo 57
E-08193 BELLATERRA BARCELONA

Prof. Dr. Kurt HELLER
Ludwig-Maximilians-Universität
Abteilung Pädagogische Psychologie
und Psychologische Diagnostik
Leopoldstr. 13
D-8000 MÜNCHEN 40

Prof. Dr. Laurence RIEBEN
Faculté de Psychologie et des Sciences
de l'Education de l'Université de Genève
24 rue Général Dufour
CH-1211 GENEVE 4

Dr. Andrzej SĘKOWSKI
Department of Psychology
University of Marie Curie-Sklodowska
Plao Litewski 5
PL-20863 LUBLIN

Prof. Dr. Pieter SPAN
Louis Bouwmeesterlaan 77
NL-3504 GE UTRECHT

II. Delegates

AUSTRIA

Magr. Barbara GRILLMAYR
Padägogisches Institut des Bundes in Salzburg
Erzabt Klotzstrasse 11
A-5020 SALZBURG

AZERBAIJAN

Dr. Vagif G. ALIKPEROV
Azerbaijan Istedad Association
Block 2, Apart. 35
Buinaksky Str. 16
370001 BAKU

BELGIUM

Prof. Dr. W. DE COSTER (excused)
University of Ghent
Psychological Institute
H. Dunantlaan 2
B-9000 GHENT

BULGARIA

Prof. Dr. Alexandra PENOVA (excused)
Sofia University
Dept. of Philosophy and Psychology
bul. Russky 15
BG-1000 SOFIA

CZECHOSLOVAKIA

Dr. Vladimir BURJAN
Ministry of Education, Youth and
Sport of the Slovak Republic
Hlboka 2
CS-81305 BRATISLAVA

Dr. Vladimir DOCKAL, Director
Research Institute of Child Psychology
and Psychopathology
Cyprichova 42
CS-81305 BRATISLAVA

Dr. Thomas KOVAC
Research Institute of Child Psychology
and Psychopathology
Cyprichova 42
CS-81305 BRATISLAVA

Dr. Antonin VRBA
Department of Mathematics
Faculty of Education - Charles University
Rettigové 4
CS-11639 PRAHA 1

DENMARK

Prof. Dr. Mogens JANSEN
The Danish Institute for Educational Research
28 Hermodsgade
DK-2200 COPENHAGEN

ESTONIA

Prof. Dr. Inge UNT
Department of Education
University of Tartu
Aardla 130-51
202400 TARTU

FINLAND

Mr. Jouni VÄLIJÄVRI
University of Jyväskylä
Department of Education
Seminaarikatu 15
SF-40100 JYVÄSKYLÄ

FRANCE

Gerard CHAUVEAU, Chercheur (absent)
Centre de recherche en éducation spécialisée
et adaptation scolaire
CRESAS - INRP
29 rue d'Ulm
F-75230 PARIS CEDEX 05

GERMANY

Dr. Heinz SCHIRP
Landesinstitut für Schule und Weiterbildung
Paradieser Weg 64
D-4770 SOEST

Ministerialrätin Mrs. Reneta MUSSO
Bundesministerium für Bildung und Wissenschaft
Heinemannstrasse 2
D-5300 BONN 2

Prof. Dr. Carl-Jürgen KALTENBORN
Allee der Kosmonauten 145
D-0-1140 BERLIN

Prof. Dr. Gerlinde MEHLORN
Fritz Siemon Str. 26/141
0-7024 LEIPZIG

Dipl.-Psych. Colleen NEITZKE
c/o The German National Foundation for Scholarship Awards
Mirbachstrasse 7
D-5300 BONN 2

ITALY

Prof. A. VISALBERGHI, President(excused)
Centro Europeo dell'Educazione (CEDE)
University of Rome
Villa Falconieri
I-00044 FRASCATI (Roma)

Prof. Giovanni COLOMBO, Secretary General
I.R.R.S.A.E.-Lombardia
Via Leone XIII, 10
I-20145 MILANO

LATVIA

Dr. Anita CAURE (excused)
University of Riga, Faculty of Education
M. Stacijas Street 20, Apart. 37
RIGA

NETHERLANDS

Dr. A. DUDINK
University of Amsterdam
Department of Psychology
Roeterstraat 15
NL-1018 WB AMSTERDAM

Drs. E. J. J. VAN HEZIK
Ministry of Education and Science
Department BO/V
P.O. Box 25000
NL-2700 LZ ZOETERMEER

PORTUGAL

Mrs. Manuela MACHADO
Ministry of Education
Directorate General for Basic and Secondary Education
International Relations Department
Av. 5 de Outubro, 107
P-1051 LISBOA 1-CODEX

ROMANIA

Hilda Minodora BURCIN
Bibliothèque Nationale Pédagogique
Strada Zalomit 12
RO-70714 BUCURESTI

Dr. Carmen CRETU
University "ALICUZA"
Department of Education and Psychology
B-del Copou, 11
R0-6600 IASI

RUSSIA

Dr. Irina AVERINA
Alabjana 12-55
125080 MOSCOW

Dr. Helena SCHEBLANOVA
Dubrovskaja 10/8 Fl. 30
MOSCOW

Dr. Yuri P. MASHAROV
Creative Union "Luchik"
PO BOX 91
630090 NOVOSIBIRSK 90

SLOVENIA

Mrs. Marjeta CERAR (excused)
Institute for Educational Research
University of Ljubljana
Gerbiceva 62
61000 LJUBLJANA

Prof. Dr. Norbert JAUSOVEC
Faculty of Pedagogics
University of Maribor
Koroska C. 160
62000 MARIBOR

SPAIN

Prof. Dr. Candido GENOVARD (excused)
Universidad Autonoma de Barcelona
Aptdo 57
E-08193 BELLATERRA BARCELONA

Dr. Javier BERCHÉ CRUZ
C.R.E.D.E.Y.T.A
Aribau 184, 1° 3°
E-08036 BARCELONA

SWITZERLAND

Dr. Hans STRICKER
Hangweg 2
CH-3047 BREMGARTEN/BERN

TURKEY

Dr. Füsun AKARSU
Inanç Vakfi
Sarayardh Cad. No. 24
TR-81010 AUBADEM-ISTANBUL

UNITED KINGDOM

Mr. D. C. McINTOSH
32 Crafts Way
SOUTHWELL
GB-NOTTINGHAMSHIRE NG25 OBL

Dr. Rosemary AYLES (excused)
University of Reading
Faculty of Education and Community Studies
Bulmershe Court
GB-EARLEY, READING RG6 lHY

YUGOSLAVIA

Ms. Slavica MAKSIÇ
Institute for Educational Research Belgrade
Dobrinjska St. ll/III
YU-llOOO BEOGRAD

Mrs. Hasnija MURATOVIC (absent)
Republicki pedagoski zavod
0. Kersovanija 3
YU-71000 SARAJEVO

III. Parliamentary Assembly of the Council of Europe

M. Xavier HUNAULT
Membre de la Commission de la Culture et de l'Education de l'Assemblée Parlementaire du Conseil de l'Europe
Documentation étrangère
Assemblée Nationale
F-75355 PARIS 07 S.P.

IV. Observers

UNESCO

Mr. Toshio OHSAKO
International Bureau of Education
Case postale 199
CH-1211 GENEVA 20

OECD (excused)

EUROPEAN COMMUNITY (excused)

WORLD COUNCIL FOR GIFTED AND TALENTED CHILDREN (WCGT)

Dr. Norah MAIER, President WCGT
University of Toronto Schools
The Faculty of Education
371 Bloor Street West
CDN-TORONTO M5S 2R8 / Canada

EUROPEAN COUNCIL FOR HIGH ABILITY (ECHA)

Paul ANDREWS, Executive Secretary (excused)
Bildung und Begabung
Wissenschaftszentrum
Ahrstrasse 45
D-5300 BONN

EUROTALENT

Louis NAZARETH, Président EUROTALENT (excused)
R. Jornal Noticias No. 207, 2°-D[to]
P-4100 PORTO

EUROTALENT
SOCIETE FRANCAISE DE PSYCHOLOGIE

Dr. Jean BRUNAULT, Président-Fondateur
EUROTALENT
15 rue Barillet Deschamps
F-37000 TOURS

Delegate of MENSA-France to EUROTALENT

Gilles DESCHAMPS (excused)
8 rue Léon Gaumont
F-30900 NIMES

ASSOCIATION NATIONALE POUR LES ENFANTS INTELLECTUELLEMENT PRÉCOCES (ANPEIP)

Jean-Charles TERRASSIER, Président
ANPEIP
366 avenue de Fabron
F-06200 NICE

EUROTALENT
ASOCIACION PARA EL DESARROLO DE LA CREATIVIDAD Y TALENTO (CREDEYTA)

Dr. Javier BERCHÉ CRUZ
Aribau, 184, 1°, 3°
E-08036 BARCELONA

NATIONAL ASSOCIATION FOR CURRICULUM ENRICHMENT & EXTENSION (NACE)

Mrs. Johanna RAFFAN, Chairman
National Association for Curriculum Enrichment & Extension
Park Campus
Boughton Green Road
GB-NORTHAMPTON NN2 7AL

NATIONAL ASSOCIATION FOR GIFTED CHILDREN (UK)

Mr. Michael SHORT, Director
National Association for Gifted Children
Park Campus
Boughton Green Road
GB-NORTHAMPTON NN2 7AL

NATIONAL ASSOCIATION FOR GIFTED CHILDREN (USA)

Dr. Barbara CLARK (excused)
23523 Canzonet
WOODLAND HILLS
California 91367, USA

NATIONAL PARENTS ORGANIZATION OF GIFTED CHILDREN (NL) "PHAROS"

Mrs. Fabienne BLOCQ
P.O. Box 22
NL-4153 ZG BEESD

Wim RENSEN
't Hart 134
NL-2266 NK LEIDSCHENDAM

AUSTRALIA

Mrs. Maureen ROBINSON
S.C.E.C.G.S. Redlands Junior Higher School
272 Military Road
2090 CREMORNE
SYDNEY
AUSTRALIA-N.S.W.

INDONESIA

Prof. Dr. Wiwi MARAT
Fakultas Psikologi
Universitas Padjadjaran
Jl. Pojok Dago 23
BANDUNG 40135
Indonesia

Dr. Hanna WIDJAJA
Fakultas Psikologi
Universitas Padjadjaran
Jl. Pojok Dago 23
BANDUNG 40135
Indonesia

Dr. Reni AKBAR-HAWADI
University of Indonesia
Fakultas Psikologi
Kampus Baru UI
DEPOK-Indonesia

PERU

Ms. Sheyla BLUMEN
Pontificia Universidad Católica del Peru
Apartado 1761
LIMA 100/PERU

Secretariat

COUNCIL OF EUROPE

Dr. Michael VORBECK
Head of Section for EUDISED and Educational Research
BP 431 R6
F-67006 STRASBOURG CEDEX

Mme Daniele IMBERT
Section for EUDISED and Educational Research
(same address)

Ms. Marcella MÖNKS
Center for the Study of Giftedness
P.O. Box 9104
NL-6500 HE NIJMEGEN

Ms. Anke SOBIERAJ
Former trainee at the Section for
EUDISED and Educational Research

Ms. Veronique RIBBINK
Former trainee at the Section for
EUDISED and Educational Research

UNIVERSITY OF NIJMEGEN

Prof. Dr. Franz MÖNKS, Director
Center for the Study of Giftedness
University of Nijmegen
PO Box 9104
NL-6500 HE NIJMEGEN

Interpreters

Mme Patricia WARD
143 avenue Orban
B-1150 BRUXELLES

Mme Willy VISSER
72 av. du Castel, Bte 12
B-1200 BRUXELLES